INNOVATION IN MUSIC

Innovation in Music: Adjusting Perspectives brings together cutting-edge research on new innovations in the field of music production, technology, performance, and business. With contributions from a host of well-respected researchers and practitioners, this volume provides crucial coverage on the relationship between innovation and rebellion.

Including chapters on generative AI, gender equality, live music, quantisation, and composition, this book is recommended reading for music industry researchers working in a range of fields, as well as professionals interested in industry innovations.

Jan-Olof Gullö is Professor in Music Production at the Royal College of Music, Stockholm, Sweden and Visiting Professor at Linnaeus University.

Russ Hepworth-Sawyer is a mastering engineer with MOTTOsound, an Associate Professor at York St John University, and the managing editor of the *Perspectives on Music Production* series for Routledge.

Dave Hook is an Associate Professor in Music at Edinburgh Napier University. A rapper, poet, songwriter, and music producer, his research focuses on hip-hop, rap lyricism, identity, culture, and performance, through creative practice.

Mark Marrington is an Associate Professor in Music Production at York St John University, having previously held teaching positions at Leeds College of Music and the University of Leeds. His research interests include metal music, music technology and creativity, the contemporary classical guitar, and twentieth-century British classical music, and his recently published book, *Recording the Classical Guitar* (2021), won the 2022 ARSC Award for Excellence in Historical Recorded Sound Research (Classical Music).

Justin Paterson is Professor of Music Production at London College of Music, University of West London, UK. He has numerous research publications as author and editor. His research interests include haptics, 3-D audio, and interactive music, fields that he has investigated over a number of funded projects. He is also an active music producer and composer; his latest album (with Robert Sholl), *Les ombres du Fantôme*, was released in 2023 on Metier Records.

Rob Toulson is Director of RT60 Ltd, who develop innovative music applications for mobile platforms. He was formerly Professor of Creative Industries at the University of Westminster and Director of the CoDE Research Institute at Anglia Ruskin University. Rob is an author and editor of many books and articles, including *Drum Sound and Drum Tuning*, published by Routledge in 2021.

Perspectives on Music Production
Series Editors: Russ Hepworth-Sawyer, *York St John University, UK,*
Jay Hodgson, *Western University, Ontario, Canada,* and **Mark Marrington,**
York St John University, UK

This series collects detailed and experientially informed considerations of record production from a multitude of perspectives, by authors working in a wide array of academic, creative and professional contexts. We solicit the perspectives of scholars of every disciplinary stripe, alongside recordists and recording musicians themselves, to provide a fully comprehensive analytic point-of-view on each component stage of music production. Each volume in the series thus focuses directly on a distinct stage of music production, from pre-production through recording (audio engineering), mixing, mastering, to marketing and promotions.

Reimagining Sample-based Hip Hop
Making Records within Records
Michail Exarchos

Remastering Music and Cultural Heritage
Case Studies from Iconic Original Recordings to Modern Remasters
Stephen Bruel

Innovation in Music
Adjusting Perspectives
Edited by Jan-Olof Gullö, Russ Hepworth-Sawyer, Dave Hook, Mark Marrington, Justin Paterson, and Rob Toulson

Innovation in Music
Innovation Pathways
Edited by Jan-Olof Gullö, Russ Hepworth-Sawyer, Dave Hook, Mark Marrington, Justin Paterson, and Rob Toulson

For more information about this series, please visit: www.routledge.com/Perspectives-on-Music-Production/book-series/POMP

INNOVATION IN MUSIC

Adjusting Perspectives

Edited by Jan-Olof Gullö, Russ Hepworth-Sawyer, Dave Hook,
Mark Marrington, Justin Paterson, and Rob Toulson

Routledge
Taylor & Francis Group

LONDON AND NEW YORK

Designed cover image: Rachel Bolton

First published 2025
by Routledge
4 Park Square, Milton Park, Abingdon, Oxon OX14 4RN

and by Routledge
605 Third Avenue, New York, NY 10158

Routledge is an imprint of the Taylor & Francis Group, an informa business

British Library Cataloguing-in-Publication Data
A catalogue record for this book is available from the British Library

Library of Congress Cataloging-in-Publication Data
Names: Gullö, Jan-Olof, editor. | Hepworth-Sawyer, Russ, editor. |
Hook, Dave, editor. | Marrington, Mark, editor. | Paterson, Justin, editor. | Toulson, Rob, editor.
Title: Innovation in music : adjusting perspectives / edited by Jan-Olof Gullö,
Russ Hepworth-Sawyer, Dave Hook, Mark Marrington, Justin Paterson, and Rob Toulson.
Description: Abingdon, Oxon ; New York : Routledge, 2024. |
Series: Perspectives on music production |
Includes bibliographical references and index. |
Identifiers: LCCN 2024027687 (print) | LCCN 2024027688 (ebook) |
ISBN 9781032500256 (hardback) | ISBN 9781032500249 (paperback) |
ISBN 9781003396550 (ebook)
Subjects: LCSH: Music–21st century–History and criticism. |
Music and technology. | Music–Production and direction. |
Artificial intelligence–Musical applications. |
Composition (Music) | Streaming audio. | Music and the Internet.
Classification: LCC ML55 .I566 2024 (print) | LCC ML55 (ebook) |
DDC 780.9/05–dc23/eng/20240906
LC record available at https://lccn.loc.gov/2024027687
LC ebook record available at https://lccn.loc.gov/2024027688

ISBN: 978-1-032-50025-6 (hbk)
ISBN: 978-1-032-50024-9 (pbk)
ISBN: 978-1-003-39655-0 (ebk)

DOI: 10.4324/9781003396550

Typeset in Times New Roman
by Newgen Publishing UK

CONTENTS

PREFACE

'You're not supposed to do that'. A phrase that is both a scolding and a drawing of the boundaries within which we are supposed to operate. It is the epitome of the socio-cultural boxes that we both find and put ourselves in, in order to make sense of the world. Boxes which, as soon as they are created, serve to limit or constrain. History is full of people who don't do what they were supposed to do; who take the (signal) path less travelled; who strike out into the unknown; who ask in response – 'why not?'. Innovation comes from challenging existing paradigms; from saying 'what if we tried this?'; looking with fresh eyes and being willing to throw away the established rules.

The Innovation in Music conference 2023 was held at Edinburgh Napier University, from the 30 June to the 2 July. The conference theme – 'You're not supposed to do that' – saw the coming together of scholars, practitioners, community activists and performers to question the current way of doing things, proposing new, ground-breaking approaches, imagined alternative perspectives, and futures outside the margins of traditional practices. This two-volume collection comprises the published proceedings from the 2023 conference. Book 1, *Innovation in Music: Adjusting Perspectives*, focuses on music, culture and society – interrogating, challenging and proposing new routes through, journeys towards, and viewpoints around the musical world. In Book 2, *Innovation in Music: Innovation Pathways*, the focus turns to reimagining, repositioning and revisiting perspectives in music production and music technology.

Book 1 opens with a written version of Nick Prior's captivating keynote, *On Error, Accident and Contingency in Music*. Here the author challenges current perspectives on intention and visibility in music scholarship, focusing on the significance of mistakes, breakdowns and accidents in music creation that have historically been under-analysed.

Following this, Chapters 2 and 3 consider elements of music streaming in contemporary society. Firstly, Matthew Lovett's *The Streaming Curve: Streaming, The S curve and Super-abundance*, maps the virological 'S curve' onto the development of music streaming, in order to better understand its past, present and future trajectory. Chapter 3 – *Missed or Postponed Innovation? The Rise and Fall (and Rise?) of Live Streamed Music Events in Italy*, then, sees Francesco D'Amato consider streaming of live music events in Italy, their popularity over time and the socio-cultural and technological levers that impact this.

AI is the focus of Rachael Drury's chapter, with generative music, AI learning and concerns around artist's rights considered in *You're Not Supposed to Launder My Music! Music as Data in the Training of Generative AI Music Models*. The future of music is also the focal point of Katy H Weatherly's chapter, in this case from the perspective of classical instrumental tuition. Her chapter, *"Unconscious Rebelliousness": The Polytonal Roles of Young Musicians in an Innovative Music Program*, considers the Face the Music programme in New York City and its hybridisation of traditional and contemporary music practices.

In Chapter 6, Corinna Woolmer investigates a very pertinent and timely issue – gender balance and equality in the music industry. *Innovation in Music: Reimagining Approaches to Gender Equality in the British Live Music Industry*, takes an interdisciplinary approach to the topic, drawing from feminism, social studies and other perspectives to provide a more rounded view of the issues, and highlighting methods for creating meaningful change. Clara Colotti's chapter, *Modes of Engagement with Classical Music: Digital Formats*, then considers the role that digital performances can play in continuing and building audience engagement for orchestral performance.

Agata Kubiak-Kenworthy's work considers the future suitability of chord charts for representing contemporary harmony in Jazz. *When Chord Charts Fail: Pitfalls of Radical Reharmonisation of Jazz Standards* presents a case study for the analysis of chord chart creation and interpretation. This is followed by a different type of case study in Jamie Birkett's *Digital Aesthetics and Transcending Lo-Fi in Alex G's 'God Save the Animals'*. Here Birkett examines the aesthetics and intertextuality resulting from the artist's choice to move from a traditionally lo-fi music production approach to a larger scale high-fidelity project using professional recording studios and engineers.

Both Chapters 10 and 11 consider music in relation to physical and cultural spaces, with Beth Karp's *Space and Place – Outsiders Collecting, Curating and Sharing Insider Stories and Sounds*, chronicling the creation of a series of musical pieces representing the communities on the Isle of Lewis & Harris of the west coast of Scotland. This is discussed alongside notions of outsiderdom, cultural belonging, heritage and creative practice. In Chapter 11, Bob Birch considers aural-architecture and site-specific composition. *Aural Architecture: Integrating Site into Composition* discusses the ways that a performance space can be connected to the composition, transforming both the piece and the space through their interconnectivity.

Hip-hop and both its local and global cultural impact are at the centre of Chapters 12 and 13. Firstly, Sace Lockhart's, *"We Went From 'Yes, Yes Yawl', Tae 'Who You Talkin Tae?'"*: *Language and Authenticity in Scots Hip Hop* reflects on the cultural capital of language and authenticity, through his analysis of early adoption and integration of hip-hop culture in Scotland. Following this, Zachary Diaz discusses the late, great J Dilla and others' contributions to the use of quantisation for groove and feel in *Might as Well be Swing: On the Use and Misuse of Quantisation in Hip-Hop Production*.

Diversity of social identity is at the forefront for Euan Pattie's study of dance music listeners' perspectives in *Innovation in Dance Music Research – a Focus on Listening*. Kirsten Hermes' chapter also centres around identity, in this case in relation to performance avatars and virtual characters. *Creative Cyborgs: Virtual 3D Characters as Artist Identities for Musicians* discusses the creation of digital avatars for virtual performance.

In Chapter 16, Corin Anderson's *Translating Artworks into Music: Synaesthetic Reverse-engineering in Music Composition* Discusses his own audio-visual synaesthesia and discusses utilising visual stimulus to create musical pieces. Extending discussions on composition, in Chapter 17, Robert Wilsmore re-envisions concepts of copyright, collaboration and creativity in *Reframing Conflicts Between Systematic Production, Creative Production, Authorship and Ownership*.

Finally, David Thyrén and Jan-Olof Gullö complete the first volume with *An Innovative Music Production Model Leading to a Sustainable Hit Song – "Främling"*. This analysis of the Swedish hit song, "Främling" by Carola Häggkvist, asks what aspects of the song's construction and release can be identified as contributing to its success, allowing a framework to emerge for the creation of a hit song.

A heartfelt thank-you to all of the contributors, authors, presenters, performers and delegates for their input both to the conference and to the chapters of this book. These works comprise a spectral snapshot of the diverse, daring and distinctive counter-perspectives for change and innovation, revision, revolution and disruption in the future of music production, performance, industry, culture and society. Keep doing what you're not supposed to.

Dave Hook (Glasgow, Scotland), February 2024

1

ON ERROR, ACCIDENT AND CONTINGENCY IN MUSIC

Nick Prior

1. Introduction

A number of disciplinary turns have put small, contingent and unpredictable phenomena in the frame. In the case of the complexity turn in the natural and social sciences, causation is seen as multiple and dependent on forces traditionally seen as unimportant, like the infamous butterfly that flaps its wings in Tokyo and causes a hurricane in Brazil (Urry, 2005). In the case of the new materialism, even small things matter because they are lively (Miller, 2005), while the micro-sociological focus on the everyday such as turn taking in conversation has been a mainstay of the discipline for decades (Drew and Heritage, 1993). The focus on things once considered insignificant or even anomalous provides opportunities to give power back to the unusual and multiple. In Science and Technology Studies (STS) this is captured in Bruno Latour's classic essay on the humble door closer in which are enrolled a bunch of delegated powers and agents, not least the porter (Johnson, 1988).

Attention to these phenomena provides three crucial insights: first, it gives a more complete picture of all the things (human and nonhuman) that are gathered when action takes place (Latour, 2005). Second, it shows that this action is never a simple matter of direct human intentionality, where what is important is the will that humans exert over an inert nonhuman world. Third, it gives analytical space to the constitutive effects of contingency, including stuff that goes "wrong". Not to dismiss moments of breakdown as trivial blemishes, deviations from the normal state of things, but to see them as permanent features of modern life. From nuclear disasters to stubbing your toe on the bed, contingent phenomena deserve proper attention.

In the following, I'm going to be dealing with a cluster of phenomena variously termed error, accident, contingency and failure in music. I'm interested in these phenomena partly because they are often ignored *as* music and *as* production, but also because they reveal more about music as a complex form of practice than conventional narratives emphasizing successful and purposeful action. The key argument I want to make is that studying accident, error and contingency makes no sense if we reduce them to exceptions to the normal state of action. Not only are these phenomena constant and abundant, but they are also constitutive. They are integral to the ordering of social, technical and musical relations. Recognizing the constitutive role of accident and contingency allows us to expand the terrain of disciplines like musicology, popular music studies and the sociology of music, as well as reconstitute the domain of the accident in histories of music that

DOI: 10.4324/9781003396550-1

are too often written as smooth, linear and devoid of the messiness of everyday practice (Baker *et al.*, 2018). It's to move away from an idealized notion of musicality to register the fundamental contingency of agency and objects.

2. Preliminary Definitions

But what do we mean by accident, error and contingency? Philosophy has its own definitions. For Aristotle, the accidental qualities of an object are opposed to its substantive essence in that the accidental represents perceptible qualities such as colour, tone, size, and shape. Hence, an object like a cat is the essence, but the fact that it's black is the accidental (Annas, 1977). Contemporary definitions offer additional lines of distinction: whilst contingency involves unexpected degrees of chance and uncertainty, accident is an unplanned event; and whilst error is usually seen as an unintentional deviation from accuracy, failure hints at the inability to hold and perform a normal state or function. To fail is to wander or to stray, even if that failure is considered essential to learning, as in the notion of "failing forward" (Maxwell, 2000).

A more sociological definition adds to these distinctions the insight that everyday situations are often bounded by social rules which, when followed, generate boundaries, worlds and frames of experience. When these rules break down or are transgressed the frame of experience is broken and work is required to re-establish the frame (Goffman, 1974). There's a whole tradition of ethnomethodological experiments (known as "breaching experiments") that wilfully disrupt these conventions in order to show the taken-for-granted rules that hold interaction together (Garfinkel, 1991). Breaking frames can often generate confusion and annoyance leading to the need for "repair work" – work that attempts to re-establish the frame. I will return to the importance of this work later.

So, we are in the realm of *degrees* of unplanned action. In fact, in some respects, what unites these phenomena is how they are represented in popular discourses as unintentional and exceptional, they are the "other" to successful and planned events. In that sense, they are often seen as unwelcome, at least from the perspective of Western instrumental rationality. To take one of these phenomena, the accident. The accident is the flipside to progress and represents a failure to control or predict. In critical safety systems, for instance, planning for contingency and accident is a way of staving off catastrophe. To speak of them is to evoke discourses of deficiency and imperfection. Something has "gone wrong" (Perrow, 1982).

Unsurprisingly, Science and Technology Studies (STS) has a lot to say about accidents because they are where the big questions of power, the unintended consequences of intended action, safety vs profit, and the nature of complex technical systems, can be observed. In one classic statement on the changing nature of modernity in the 1990s, the social theorist Ulrich Beck argued that we now inhabit a risk society in which "the unknown and unintended consequences come to be the dominant force in history and in society" (Beck, 1992: 22). That's because increasing levels of global interconnection and interdependence increase our vulnerability to boundaryless outcomes: pollution, radioactivity, climate change, and viruses. There's also a small but critical *humanist* strand of inquiry into accidents associated with the urbanist Paul Virilio. For Virilio (2007), to interrogate the accident is to show, following Aristotle, how the particular characteristics of an entity reveal its substance or essence. However, Virilio reverses this formulation because the invention of any substance is at one and the same time the invention of a distinct accident associated with its essence. Hence, just as the shipwreck is the by-product of the ship, so the air disaster is the hidden production of the airliner and electrocution the by-product of electricity. The tragedy of modernity, for Virilio, resides in its progressive inability to tame progress, resulting in the mass production of accidents.

3. Reinstating Failure

Might Virilio's provocative thought be directed at music technologies? Is it inevitable, for instance, that the invention of the turntable is also the invention of its breakdown and how does this differ from it being repurposed to scratch vinyl? Is the inescapable fate of the guitar the fact of it being smashed, distorted and broken? And what about the lines of code that constitute the world of VSTs and DAWs? How does that "break" and get repaired?

Before I offer some headway with these questions, let me anticipate a rebuttal. Why do we need to re-instate failure when it already has an internal life within the domain of music itself? After all, pop memorabilia and heritage are full of tales of misfortune, mishap and catastrophe. Just to take one case: stories of dramatic rock star lives and deaths are in many ways a core discursive strand that hold the myth of rock together. Jeremy Simmonds' (2008) weighty *Encylopedia of Dead Rock Stars* is a good example. Ordered chronologically, Simmonds catalogues the deaths of rock stars deemed "unusual" or "extraordinary". Each entry is allocated a set of symbols to categorize various causes of death amongst rock's pantheon of stars: from car crashes to electrocutions, stage deaths to a category he calls "accidental deaths". We could also argue that whole swathes of music are already inspired by serendipity, happenstance and failure. Popular music is heavily indebted to such phenomena as fertile sources for its own musical content, after all. Songs, lyrics and whole genres (such the Blues and Country and Western) owe themselves to the intensities and absurdities of everyday breakdowns (Foster, 2022). Pop's history bears the marks of failure right through its core, in the forms of topic, trope and tragedy because failure is a particularly effective theme through which life stories and identities are given shape (Frith, 1996). From romantic breakups to political dislocation, urban decay to personal awkwardness, failure is an ever-present in pop's own repertoires.

All of this is true, but the argument I want to make is that the domain of the accidental is both more ordinary and more variant than these reflections imply. For a start, it makes no sense to reduce the domain of the accident to exceptions to the normal state of things. Nor is it useful to attach discourses of contingency to ideologies of excess, romanticism or genius as in the case of dead rock stars. Not only are contingent phenomena an ever-present, but they give shape to everyday social, technical and musical relations (White, 1992). So, whilst it remains the case that musicians have direct experience of the mess and muddle of practice, academic studies (with some notable exceptions) rarely treat this messiness seriously. Textbook accounts of contemporary popular music, for instance, are often written as relatively smooth, linear narratives devoid of the complexities of vernacular creativity, while historical overviews of pop tend to be laid out in a progressive, genre-by-genre narrative that deal with the spectacular and successful: music as a series of conventions, styles, and industry structures, for instance (Hoffman, 2015).

4. A Typology

If we *are* relatively sympathetic with this attempt to stick up for the contingent, how do we begin to plot all this stuff: what are the parameters? The following comprises an attempt to construct a dual-axis schema. One axis covers degree of visibility – how manifest, noticed or public the failure or contingency is. The other axis deals with intentionality, defined by the presence or not of conscious intent or motivation on the part of the agent. The nature of intentionality, the agent and agency are themselves matters of some debate and speculation, of course, and this needs proper consideration on an ontological level (Vandenberghe, 2002). But, the purpose is not to come up with a definitive set of pronouncements that exhaust the nature of these practices, but to compile a suggestive set of criteria that opens an analytical space from which further refinements can be made (Figure 1.1).

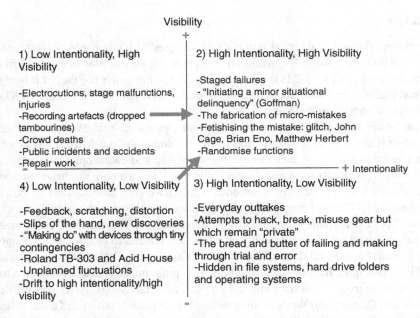

Visibility

1) Low Intentionality, High Visibility

-Electrocutions, stage malfunctions, injuries
-Recording artefacts (dropped tambourines)
-Crowd deaths
-Public incidents and accidents
-Repair work

2) High Intentionality, High Visibility

-Staged failures
- "Initiating a minor situational delinquency" (Goffman)
-The fabrication of micro-mistakes
-Fetishising the mistake: glitch, John Cage, Brian Eno, Matthew Herbert
-Randomise functions

Intentionality

4) Low Intentionality, Low Visibility

-Feedback, scratching, distortion
-Slips of the hand, new discoveries
- "Making do" with devices through tiny contingencies
-Roland TB-303 and Acid House
-Unplanned fluctuations
-Drift to high intentionality/high visibility

3) High Intentionality, Low Visibility

-Everyday outtakes
-Attempts to hack, break, misuse gear but which remain "private"
-The bread and butter of failing and making through trial and error
-Hidden in file systems, hard drive folders and operating systems

FIGURE 1.1 A provisional typology of contingency in music.

The schema splits into four quadrants, with intentionality running on the x axis and visibility along the y axis, and it generates the following categorizations.

4.1. Quadrant 1: Low Intentionality, High Visibility

In the top left quadrant, we might recognize those spectacular phenomena that have low intentionality but high visibility like electrocutions, exploding equipment, stage malfunctions, broken strings, injury, and mishaps. Visibility is partly a function of how well known a performer is, of course, so when Michael Jackson's hair catches fire during the shooting of a Pepsi commercial in 1984, this is public precisely because of Jackson's position in the field of popular music. It's also quite difficult to separate fact from fiction, here. For instance, Iggy Pop once claimed he had fallen off every stage he had ever been on. But hopefully, the idea is clear enough. This quadrant covers incidents that are visible, that gain a high degree of public exposure, but where agency (profoundly distributed as it is) is opaque and contingency high.

Other examples might be misprinted record sleeves (which often end up as collector's items) and unintentional mistakes or imperfections made in the recording studio that end up as celebrated and recognized parts of the recording itself. The Beatles' songs are littered with such examples – the sound of a dropped tambourine during the song "I'm Looking Through You" being a good example. In a similar vein, Thom Yorke of Radiohead remembers the recording of "Fake Plastic Trees" as "a fucking nightmare" because at "one stage at the first session . . . it sounded like Guns N' Roses". Yorke explains that the song was saved through a mixing error: "Paul [Kolderie] missed a cue, so the electric guitars don't come in at the right place. It was a mistake, but we kept it" (Randall, 2000: 114). Famously, the iconic gated reverb drums on Phil Collins' "In the Air Tonight" are, by all accounts, a result of serendipitous factors in the sound engineer, Hugh Padgham's, handling of the talkback feature of the mixing desk (Buskin, 1992), while also included here would be tragic instances of crowd death such as at the Roskilde Festival in 2000 when nine fans were crushed to death.

As noted earlier, from a sociological perspective, it's interesting to note how moments like these often give rise to what ethnomethodologists call "repair work" – that is, ongoing action that attempts to apologize for, rectify or account for the incident (Denis, 2020). This kind of repair work is sometimes so elaborate and ritualized that it outdoes the original mistake in terms of its importance to the resulting encounter. I once interviewed the coordinator of the live events of the Japanese virtual star Hatsune Miku, and he told me that in one live performance, the Miku hologram stopped working. The banks of computers which generated the singing avatar and the backup computers both broke down, leaving the gig in limbo. But rather than complaining or booing, the audience clapped in support of the technicians and shouted "*ganbatte*", Japanese for "do your best". When the gig finally re-started there was an extra energy and sense of camaraderie in the crowd. The breakdown produced an upsurge in goodwill as the audience recognized the problem and difficulties with the technology. They become the repairers of the gig as a social encounter (Prior, 2021).

4.2. Quadrant 2 (High Intentionality, High Visibility)

As we move to the second quadrant of high intentionality and high visibility, we can recognize that at times, especially in the live context, some performers purposely disrupt their own performances in order to play with or break the frame of performance itself. Staged failures and mishaps have a long history in performance because they create the circumstances in which audiences feel more involved or sympathetic with the performer. "Initiating a minor situational delinquency" as the sociologist Erving Goffman puts it (Goffman, 1974: 423) helps to generate extra frisson and a sense of spontaneity in the moment. Here, contingency can be actively fabricated as a way of signifying liveness and human presence. While we certainly don't want the trapeze artist to fall and break their neck, it is the wobble that generates the thrill and it's often said that James Brown used to fabricate little micro-mistakes during live performances to keep the audience interested (Frith, 1996).

Also in this quadrant, but in a more avant-garde register, is the rich tradition of fetishizing the indeterminate. John Cage's experiments in aleatory music, Brian Eno's oblique strategies cards and Matthew Herbert's *Manifesto of Mistakes* are modern examples of musicians orienting their work practices to be *open* to the idea of contingency as a kind of aesthetic formula (Albiez and Pattie, 2016). This is also the case with genres like glitch and a good proportion of contemporary electro-acoustic music that utilizes chance and error in the process of making and performing. In short, this quadrant shows how contingency can be wilfully captured as a resource for musicians' own interventions and position-takings. There's even an industry based around replicating these errors: software plug-ins that produce random glitches to order, pre-recorded sounds of vinyl static or "off-grid" rhythms (Danielson and Brøvig-Hanssen, 2016). Over the last couple of decades, we've seen the advent of commercially available software plug-ins with in-built randomizers that use algorithms to randomly generate sounds, patterns or sequences. The same, increasingly, with hardware.

Eventually situated at the intentional end of the spectrum, this is contingency reframed and appropriated – as reified and commodified. Intentional slips can turn the mistake into a spectacle and the resulting dynamic of tension/release can become a constitutive part of musical conventions and how to package them. Error, it seems, can sell.

4.3. Quadrant 3 (High Intentionality, Low Visibility)

At first sight, there's less to be said about this quadrant. It would include those incidents that are similar in levels of intentionality to the upper right quadrant (high intentionality, high visibility), but which have a more secret life. Amateur bedroom musicians forcing their gear to do unexpected things in private, or the conscious manipulation of the sounds of error that never quite materialize or that remain dogged by practical problems, for instance. Again, these might be wilfully appropriated contingencies, but which have low degrees of exposure or visibility: they stay put. Everyday outtakes where musicians are consciously trying to hack, break or misuse gear but which never turn into anything significant would be another example.

And yet, in many respects, this is the everyday experience of most musicians: the bread and butter of failing and making. It generates material that resides in and often remains on hard drives and cloud storage systems, hidden in folders and file systems. This poses huge methodological questions in terms of writing music histories. There is a whole underbelly of music that has a secret life, a whole corpus of cultural production that is never considered *as* production: neglected takes, outtakes and mis-takes, partially finished tracks, under-developed snippets, songs that never made the cut, unmixed tracks, fumbled recordings, and so on. It's not that these materials have been completely ignored, of course. After all, the image of the discarded tape reel is, in many ways, a stock image. It hints at those undiscovered classics and secret recordings that reveal more about one's heroes than the polished studio product. In the main, however, the discarded remains so. This is particularly the case in the digital age, where the digital detritus of outtakes and mistakes has less of an overtly material life, hidden by complex file systems, undo strokes and ever-expanding storage media.

The management and disposal of musical files articulates a tension between absence and presence, transience and permanence and between past and present. Here, we can draw upon post-structuralist writings on waste to make sense of these dynamics. Disposal, from a post-structuralist perspective, is never a final act of rendering invisible or making inert. Rubbish never disappears for good in an act of total closure. "Getting rid of something is never simply an act of waste disposal", as Hetherington puts it (Hetherington, 2004: 159). Disposal instead is an open-ended or recursive process because discarded material has a tendency to return: antique furniture, tat at car-boot sales, things of sentimental value, discarded objects that leak out of cupboards. Discarded objects can also transform storage systems such as attics, tips, garages, fridges, wardrobes, make-up drawers and basements, as well as our sense and memories of the original objects themselves.

With respect to music, tapes, CDs, digital audio or MIDI files aren't consumer goods in the usual sense. They aren't made rubbish in quite the same way, partly because they are bound up with emotional and creative labour. But there is a similarity. We hear of stories of discarded tapes of famous musicians being saved from the bin, rare cuts being found in attics, newly found archives, vinyl taking up too much room in domestic spaces, and so on. In 2008, unsold CDs of Robbie Williams' album *Rudebox* even turned up as road-surfacing material in China.

In digital systems, it is the folder, the desktop, the window and the trash bin that represent dynamics of disposal, and these interfaces hint at how we might understand the digital archive of music, including its open-endedness. At a mundane level, for instance, we might explore the ways in which old takes return in new material. One Edinburgh-based electronic musician put it to me as follows:

> When recording live instruments, like say, if I've recorded live drums and I've got lots of takes and say there is something missing from the take that I've saved, I'll go back to old ones, which are discarded and look for a particular snare sound that I want or whatever.
>
> *(Respondent A).*

In another interview, a respondent reveals a logic (perhaps even an ethics) of digital recycling and reusing:

> Nothing's ever wasted . . . if an idea doesn't come, if it's not really working, it gets shelved for a bit, but it's more than likely it'll get mined again or perhaps it gets sent on to another band member who uses an MPC and he'll chop it up and it'll become something else.
>
> *(Respondent B).*

A further respondent points to a more complex relationship to the memory and process of loss, including how losing digital music data can turn out to be a positive thing when he says:

> My younger brother and the lead singer of our band have lost entire . . . you know, maybe six months of work and . . . it kind of worked as a blessing in disguise for one track because he got to a stage where he wasn't sure whether or not it was finished and we needed to draw a line under it.
>
> *(Respondent C).*

A final respondent speaks of how an outtake becomes a "ghostly" presence existing between absence and presence: "it lurks there", he says (Respondent D). This, in turn, points out the recursive nature of these processes – muted but never absent. Indeed, while it's beyond the scope of this chapter, there are clear resonances with the idea of hauntology and how once discarded musical ephemera have hauntological qualities: sometimes to return, nostalgically, as "retro" (Reynolds, 2011), or as part of an uncanny sense of historical time (Fisher, 2022).

4.4.   Quadrant 4 (Low Intentionality, Low Visibility)

It's the final quadrant of low intentionality and low visibility that I want to finish on, however, because it's where the contingent uses of technology can sometimes lead to extraordinary discoveries and an eventual drift from low visibility to high visibility. Such is the case with feedback, turntable scratching and guitar distortion – the latter reputedly the result of accidents in which the cones of amplifiers and vacuum valves were first accidentally and then purposely damaged by blues musicians (Seed *et al.*, 2003). This quadrant is representative of the kinds of ordinary contingent phenomena that producers have direct experience of, such as slips of the hand, happy accidents, novel chord sequences found by accident and so on. In the course of "making do" with artefacts and devices, we might recognize how tiny contingencies can have larger systemic effects, too.

A good example is the history of the Roland TB-303 analogue synthesizer and the birth of acid house. Released in 1982 as a "set and play" bassline generator to accompany guitarists, the 303 didn't sell well and second-hand units eventually found their way into the hands of Detroit-based musicians DJ Pierre and DJ Spanky. In the guise of the band Phuture, Pierre and Spanky twiddled the resonance and cutoff knobs of the 303 in real-time to create the song Acid Trax – a track characterized by the well-known squelch of what was to become acid house (Reynolds, 1998). The tweaks made to the device sparked a series of larger developments to a whole genre, at the same time shaping everything from sub-cultural affiliations and drug culture to moral panics and legislation targeted at acid house parties and rave culture (Hill, 2002).

What is the best way to understand the case of the Roland TB-303? In a recent article, Paul Harkins and I (Harkins and Prior, 2021) explore how terms like "democratization" are insufficient in cases like this, because they gloss basic questions around affordability, class and racial inequalities. Neither are we talking about the improvisations of seasoned jazz musicians laced as they are with ideologies of seniority, mastery and virtuosity, or the naive fumblings of the jejune. In the past, I've been taken to task by Mark Fell in a piece he wrote in *The Wire* magazine about the characterization of glitch and acid house as contingent (Fell, 2013). As I was using terms like "instabilities" and "programming mistakes", he found my account to be loaded with assumptions about the anomalous nature of Phuture's actions. I think this is a reasonable criticism, and it suggests that in accounting for moments like this, we often still carry the normative baggage of error and accident as "other" to the normal state of things.

There are other issues and problems to consider: for instance, the exceptions that problematize this typology, phenomena that don't quite fit anywhere. At what point does an incident become a music related incident, for instance, as opposed to just an incident? What role does pure chance play: do chance meetings between musicians count? Greil Marcus writes of Bob Dylan's single "Like a Rolling Stone", for instance: "had circumstances been ever slightly different – different people present, a different mood in the studio, different weather in the street outside, a different headline in the morning paper – the song might never have entered time at all" (Marcus, 2005: 4). If we follow this logic, isn't all music defined by a series of chance happenings?

What makes these problems even more acute is that what often appear to be unintentional mishaps or accidents – arbitrary and devoid of causation or blame – turn out to be caused by some hidden or deeper causation. Many of the tragic accidental deaths of rock stars could presumably be reframed as the product of negligent health and safety procedures, as well as cutbacks in the culture industries and, even more broadly, the profit-first motives of capitalism. And, of course, after Freud, it's quite difficult to accept that slips and mishaps lack any kind of deeper cause. As for chance, aren't the dice always loaded by basic sociological indices like power and inequality? To be in the same room as music technologies already speaks of embedded social trajectories (such as enrolment in tertiary education, leisure time and the acquisition of technical capital) that are unevenly distributed and experienced along lines of stratification like gender, race/ethnicity and class (Born and Devine, 2015).

There's also profound historical variation in the definition and framing of these phenomena. Take the presence or not of distortion, hiss or static: for a while, in the very early years of sound recording, hiss was heard as relatively intractable; but as the 20th century proceeded, hiss became noise in the system to be overcome through various technical means, including noise reduction processes (Sterne, 2006). Nowadays, at a point when it's possible to produce extremely low-noise digital signals, hiss and other audio artefacts like vinyl crackles have made a return amidst a mix of nostalgia and post-digital resurrection of the perceived authenticity of an earlier age. Hence, rather ironically, the rise of a whole industry of digital plug-ins, tape saturation algorithms and Boards of Canada-esque treatments to make digital recordings sound aged and less digital.

In which case, might we say that accidents, errors and mistakes are contested, negotiated, unevenly distributed and historically specific? If so, what this rather static schema does is fail to represent the dynamism and ambiguity of the phenomena under study.

One move we could make is to explore how phenomena shift from one quadrant to another as technology, culture and the field of popular music shift. One pervasive drift is from the bottom left to the top right, as unintentional and less visible phenomena become wilfully appropriated and crystallize into visible conventions, genres and standardized ways of doing music. It would also be necessary to show how intentionality is a slippery and ambiguous term, sparking all sorts of permutations around consequences and action (Giddens, 1984). One of these relates to the claim that intention is never really a result of pure human volition, at all, but is a messy result of human-non-human engagements and entanglements. Musicians never music alone, after all, they are not the sole authors of their works, so perhaps we're better off talking about human intention as constantly deflected, indirect or co-extensive with the inner lives of machines, devices and objects. This would be to move to a flatter ontology associated with Actor Network Theory where the boundaries between bodies and technologies are always blurred and where nonhumans are, in their material affordances, constantly enrolled in these actions and sometimes even push back or fail (Latour, 2005). Then again, is it good enough to assign the same level and type of agency to nonhumans as to humans? Do machines "intend" anything?

There are undoubtedly many more faults with the scheme and as a flat representation, it needs to be pushed and pulled into a 3D space that recognizes gradations of the social and the musical, the human and nonhuman. But as a provisional classification at least it manages to show some of the variations in the phenomena I'm interested in, and if we keep in mind the historical drift and lack of tight boundaries, it's perhaps at least a starting point.

5. Conclusion

Let me return to the question I started with. What do we gain by re-writing an account of music as and through contingency? In this chapter, I have suggested that we gain a more detailed, realistic and extended notion of what production involves. If we take contingency seriously, it shouldn't just be successful action that filters through to our commentaries and treatments. Normality and happenstance do not belong to two ontological planes but are part of the same bundles of social action. Contingency isn't normatively "other" but ordinary, constitutive and a permanent feature of music. It is bound up with the unavoidable and recursive relations between bodies, practices and technologies. The rather startling implication is that we've traditionally ignored a whole corpus of relevant action. In popular music studies at least, the focus of attention is usually the finished product, the performance or the recording. This is bound up with a particular way of writing popular music history based on a tendency towards identifying key performances as iconic, key albums, genres and songs as influential, key acts as central to the canon, and so on. In many accounts, the focus is on the successfully produced commercial recording which has a degree (more or less) of stability and public exposure.

Exploring a more symmetrical approach to success and failure could have the effect of extending understandings of production which may, in turn, help displace romantic ideas of authorship. We get further into the problem if we twist the idea of practice away from simple, purposeful action towards something altogether more complex but ordinary – an approach to music that recognizes the role of open-ended, contingent and indeterminate practices. Re-instigating failure is not necessarily to fetishize or celebrate it, nor is it to ascribe characteristics of "otherness" – as lacking in success. Instead, it's to show how the unplanned, micrological stuff of music cultures lubricates music history – the little stuff that can become the big stuff, the moments that can become the movements. Time, then, to develop a solid interdisciplinary understanding of them.

References

Albiez, S. and Pattie, D. (eds) (2016) *Brian Eno: Oblique Music*. London: Bloomsbury.

Annas, J. (1977) 'Aristotle on Substance, Accident and Plato's Forms', *Phronesis*, 22(2), pp.146–160.

Baker, S., Strong, C., Istvandity, L. and Cantillon, Z. (eds) (2018) *The Routledge Companion to Popular Music History and Heritage*. London: Routledge.

Beck, U. (1992) *Risk Society: Towards a New Modernity*. London: Sage.

Born, G. and Devine, K. (2015) 'Music Technology, Gender, and Class: Digitization, Educational and Social Change in Britain', *Twentieth-Century Music*, 12(3), pp.135–172.

Buskin, R. (1992) 'Hugh Padgham: In the Air Tonight', *Recording Musician*, October 1992, pp.29–33.

Danielson, A. and Brøvig-Hanssen, R. (2016) *Digital Signatures: The Impact of Digitization on Popular Music Sound*. Cambridge, MA, and London: MIT Press.

Denis, D. (2020) 'Why Do Maintenance and Repair Matter?', in A. Blok, I. Farías and C. Roberts (eds) *The Routledge Companion to Actor Network Theory*. London: Routledge, pp.283–293.

Drew, P. and Heritage, J. (1993) *Talk at Work: Interaction in Institutional Settings*, Cambridge: Cambridge University Press.

Fell, M. (2013) 'Collateral Damage, *The Wire*, January 2013, available at www.thewire.co.uk/in-writing/essays/collateral-damage-mark-fell (Accessed 30th August, 2023).

Fisher, M. (2022) *Ghosts of My Life: Writings on Depression, Hauntology and Lost Futures*. Winchester and Washington: Zero Books.

Foster, B. (2022) *I Don't Like the Blues: Race, Place and The Backbeat of Black Life*. Chapel Hill, NC: North Carolina Press.

Frith, S. (1996) *Performing Rites: Evaluating Popular Music*. Oxford: Oxford University Press.

Garfinkel, H. (1991) *Studies in Ethnomethodology*. Cambridge: Polity Press.

Giddens, A. (1984) *The Constitution of Society: Outline of a Theory of Structuration*. Berkeley and Los Angeles, CA: University of California Press.

Goffman, E. (1974) *Frame Analysis: An Essay on the Organization of Experience*. Boston, MA: Northeastern University Press.

Harkins, P. and Prior, N. (2021) '(Dis)locating Democratisation: Music Technologies in Practice', *Popular Music and Society*, 44(5), pp.84–103.

Hetherington, K. (2004) 'Secondhandedness: Consumption, Disposal, and Absent Presence', *Environment and Planning D: Society and Space*, 22(1), pp.157–173.

Hill, A. (2002) 'Acid House and Thatcherism: Noise, the Mob, and the English Countryside', *British Journal of Sociology*, 53(1), pp.89–105.

Hoffman, F. (2015) *History of Popular Music: From Edison to the 21st Century*. London: Paw Paw Press.

Ihde, D. (2002) *Bodies in Technology*. Minneapolis, MN and London: University of Minnesota Press.

Johnson, J. (1988) 'Mixing Humans and Nonhumans Together: The Sociology of a Door-Closer', *Social Problems*, 35(3), pp.298–310.

Latour, B. (2005) *Reassembling the Social: An Introduction to Actor Network Theory*. Oxford: Oxford University Press.

Marcus, G. (2005) *Like a Rolling Stone: Bob Dylan at the Crossroads*. London and New York: Faber and Faber.

Maxwell, J. (2000) *Failing Forward: Turning Mistakes into Stepping Stones for Success*. New York: Harper Collins.

Miller, D. (ed.) (2005) *Materiality*. Durham and London: Duke University Press.

Perrow, C. (1982) *Normal Accidents: Living with High-Risk Technologies*. New York: Basic Books.

Prior, N. (2021) 'STS Confronts the Vocaloid: Assemblage Thinking With Hatsune Miku', in A. Hennion and C. Levaux (eds) *Rethinking Music Through Science and Technology Studies*. London: Routledge, pp.213–226.

Randall, M. (2000) *The Radiohead Story: Exit Music*. London: Omnibus Press.

Reynolds, S. (1998) *Energy Flash*. London: Picador.

Reynolds, S. (2011) *Retromania: Pop Culture's Addiction to Its Own Past*. London: Faber and Faber.

Seed, M., Shepherd, J, Horn, D., Laing, D., Oliver, P. and Wicke, P. (2003) *Continuum Encyclopedia of Popular Music of the World: Production and Performance, Vol. 2*. London and New York: Continuum.

Simmonds, J. (2008) *Encyclopedia of Dead Rock Stars*. Chicago, IL: Chicago Review Press.

Sterne, J. (2006) *The Audible Past: Cultural Origins of Sound Reproduction*. Durham and London: Duke University Press.

Urry, J. (2005) 'The Complexity Turn', *Theory, Culture & Society*, 22(5), pp.1–14.

Vandenberghe, F. (2002) 'Reconstructing Humants: A Humanist Critique of Actant-Network Theory', *Theory, Culture and Society*, 19(5), pp 51–67.

Virilio, P. (2007) *The Original Accident*. Cambridge: Polity.

White, H. (1992) Identity and Control: A Structural Theory of Social Action. New Jersey: Princeton University Press.

2

THE STREAMING CURVE

Streaming, The S Curve and Super-abundance

Matthew Lovett

1. Introduction

In the early years of the 21st century, digital streaming technologies and platforms (DSPs) set out to do things differently and disrupt long-established behaviours in music consumption, transforming how we listen to music, how we pay for music, and – to an extent – how we create music.

As of 2023, in its annual Global Annual Music Report, the International Federation of the Phonographic Industry (IFPI) reports that global revenues from music streaming increased through 2022 by 11.5% to reach US$17.5 bn (IFPI, 2023). In addition, global revenues from subscriptions to streaming services also grew by 10.3% to reach US$12.7 bn (ibid.). However, while these successes are encouraging, they must be seen in the context of an overall slowing in growth rates; against 2021 figures, where global streaming revenues had grown by 23.9% (ibid.), 2022 showed a slowdown in growth of over 50%. In addition, as Spotify's quarterly figures for 2023 show, while increases in paid subscriptions have facilitated continued overall growth that reflects global trends, ad-supported revenue has been in decline in Q1 and Q2. For Q1, ad-supported revenue was €329 mn. However, the cost of achieving that revenue was €339 mn, an operating cost of €10 mn (Spotify, 2023a). For Q2, which reported in July 2023, ad-supported revenue was €404 mn, at a cost of €427 mn, showing a loss of €23 mn (Spotify, 2023b). While the tone of the IFPI's report is upbeat, it would seem that the more nuanced picture of success that Spotify's financial statements reveal is also being acknowledged. Indeed, in her foreword, Chief Executive Frances Moore recognises that, although 'the opportunities for music continue to expand' the challenges are becoming 'increasingly complex' (Moore, in IFPI, 2023).

Similarly, in late 2022 and early 2023, MIDiA Research, the music and entertainment industry analytics company, in its predictions for 2023, foresaw a challenging year ahead in music streaming. Music analyst Mark Mulligan commented that streaming revenues for Q3 in 2022 were up by only 7% in comparison to Q3 in 2021 which had seen growth of 31%; a 26% decline and a trend which at that time looked set to continue (Mulligan, 2023). Furthermore, in its predictions report for 2023, MIDiA suggested that 2023 would mark the end of an era characterised by waves of disruption in the consumer technology sector (Cirisano et al., 2023), a clear indication that the challenges caused by declining growth figures are part of a complex environment, characterised by several intersecting economic and technological transformations.

DOI: 10.4324/9781003396550-2

Other music analysts have also reflected on the challenges facing streaming during 2023. David Turner has commented on the decline in ad-supported revenue across the streaming sector, along with staff redundancies at a number of major digital streaming platforms (DSPs) and the threat to DSPs from the non-DSP sector, in particular, TikTok (Turner, 2023a and 2023b). Stuart Dredge drew on Mark Mulligan's perspectives to further emphasise the decline in ad-supported revenue, and highlighted that, while streaming subscriptions may well remain an 'affordable luxury' for many music consumers, this may well be at the expense of the live music sector. This is to say that, in the context of the current global cost of living crisis, many music fans may be forced to prioritise their entertainment consumption and reduce their spending on live music events; a reduction that could affect brands' advertising spend, which in turn could further impact ad-based income for DSPs (Dredge, 2023).

Given these challenges to the streaming sector, in what follows, I explore some of the key challenges and complexities facing the music streaming sector, and draw on Jonas Salk's S curve paradigm as a means to frame what could be seen as an inflection point facing the accepted streaming model. As modes of music consumption change in response to a variety of technological and economic disruptions, it seems that an economy that flourished in the first decades of the 21st century by reformulating the rules for how the world engages with music might now be in danger of becoming the music industry of yesterday.

2. The 'S Curve'

In 1981, the virologist Jonas Salk, along with his son, Jonathan Salk, published the book *World Population and Human Values: A New Reality* (Salk and Salk). The book drew on the virologist's work going back to the 1950s which had analysed the trajectory of global population growth. At that time, it was commonly thought that the population would continue to increase endlessly, but Salk's analysis suggested that growth would likely slow, and population numbers would plateau (Salk, 2019), a prediction borne out by current UN analysis (United Nations, 2023). Mapped onto a graph, this period of growth, slowdown and plateau looked like the curve of a letter S. In the book, Jonas Salk developed the notion of the 'sigmoid curve' – or S curve – as a way of looking at social and economic development, proposing that societal development and progress do not always move forward in a straight line (Salk, 2019). In fact, the Salks' work demonstrates that the idea of progress is itself a myth, and that the cycles of societal development are far more complex.

Jonas Salk saw the curve could be divided at its mid-point, the point of change between a period of rapid growth and a slower, more settled period. Salk called these two periods Epoch A and Epoch B. In Epoch A, resources seem infinite and the potential to expand seems to be without limits. As a consequence, there is unrestricted growth, and exploitation of natural and human resources is rewarded. In this sense, Epoch A can be characterised by short-term thinking coupled with high rates of consumption. In Epoch B the situation changes dramatically. Resources become limited and growth slows. As a result, high value is given to 'conservation, interdependence, sustainability and long-range thinking' (ibid.). While Salk clearly viewed Epoch B as the more favourable of the two situations, he felt that the transition between them would not come about simply because people argued that B was a better way of living than A. Instead, he saw that change would only result from human instincts for self-preservation. Furthermore, although positive benefits to the development and broader evolution of human societies would be understood in the long term, in the short term, the transition would be difficult, chaotic and discordant (ibid.). As such, the S curve is a compelling model for understanding and anticipating change in society at a macro-level. Turning back to the music streaming economy, if we apply the S curve's basic principles of rapid growth, slowdown and plateau to streaming's developmental trajectory, then the image of a 'streaming curve' enables us to understand the complexities of this trajectory, and to consider the likelihood and potential characteristics of any oncoming transition.

3.   A Brief History of Music Streaming

In one form or another, music streaming has been in existence for over 25 years. Building on Sony Entertainment's explorations into streaming which had been running since 1997, two commercial streaming platforms were launched in 2001; Musicnet and Pressplay. Musicnet was backed by BMG, EMI, and Warner Music Group, while Pressplay, using rights management technology adopted from Microsoft's Windows Media Player, was backed by Universal Music Group and Sony (Turner, 2019b). The music writer and former *Emerging Artist Lead* at SoundCloud, David Turner, reports that Pressplay made royalty payments of $0.0023 per stream based on a subscription of $19.95, less than half of either Spotify's or Apple's current rate of remuneration. Perhaps unsurprisingly, this led to several musicians including Dixie Chicks, Dr Dre and No Doubt, instructing their lawyers to sue the platform, before removing their music from both Pressplay and Musicnet (ibid.). Rhapsody launched in 2002; the first platform to offer unlimited streaming for a monthly subscription fee, which helped it to gain a significant following among music enthusiasts. Following this, Pandora launched in 2005. Operating almost like a personalised radio service, rather than allowing users to select individual songs or albums, Pandora used an algorithm to create playlists for users based on their musical preferences. Founded in 2006, Spotify initially launched in 2008, before going on to launch in the US in 2011. Its founders, Daniel Ek and business partner Martin Lorentzon, had wanted to develop a legal alternative to file sharing that would enable music fans to access music, but that compensated artists and rights holders. Of the major streaming services, Amazon Music, Deezer and SoundCloud were launched in 2007, Google Play in 2011, Tidal in 2014 and Apple Music in 2015.

While the history of the music industry over the last 30 years is often characterised as being essentially a transition away from an economy driven by the sales of physical music products – in particular compact discs – towards an economy driven by digital streaming, there are complexities that can be overlooked, which can give us a much fuller understanding of the contemporary music landscape. It is true that, due to a peak in CD sales, the music industry had boomed in the late 1990s. The North American music industry has long been the largest of the global music markets, and the RIAA reports that in 1999, in the US, CDs contributed $22.5 bn to the industry's overall revenue of $25.6 bn (figures adjusted for inflation against the value of the dollar in 2022) (RIAA, 2023). This peak was followed by an overall decline across the music industry that continued through to 2014–15, with the US market sinking to $8.3 bn (adjusted for inflation against 2022 US dollars) (ibid.). Against this overall decline, digital streaming began to grow from 2005 onwards, initially via paid subscriptions and then ad-supported options from 2011, to eventually reach current overall levels for streaming of $11.2 bn (adjusted for inflation against 2022 US dollars) (ibid.).

On the surface, these figures present a reasonably straightforward picture of an industry in transition. However, David Turner has gone further, suggesting that, ultimately, the transition is being driven by an overarching financialisation of the music industry, a perspective he shares with the writer and label founder Gabriel Meier. Turner charts the preponderance of record label mergers and acquisitions in the 1980s and 1990s, citing the purchase in 1979 of EMI by Thorn Electrical Industries Limited, a British electronics company, as the first example of a music company being added to the portfolio of a previously non-music corporation (Turner, 2022a), and therefore a purchase driven by apparently purely financial, rather than musical concerns.

This pattern of acquisition continued with Sony's purchase of CBS in 1987, following the label's failure to capitalise on the success of some of its core artists, including Michael Jackson, Bruce Springsteen and Bob Dylan (ibid.), and, in 2004, Bain Capital led a consortium to purchase Warner Music Group, resulting in 2000 redundancies from a 6500 workforce, a $250 mn reduction in the label's operating budget, and $350 mn being paid out to the new investors (Meier, 2022). For Meier, the challenges that the music industry faced in the early 2000s which gave rise to music streaming, followed on directly from this period of financialisation, with a 'second wave' that was characterised by a series of 'mass layoffs, record store closures and artist-roster cuts' that were ultimately focused on 'overall profitability' (ibid.). This volatility in music reflected a wave of disruptions sweeping through the digital economy at this time, where, after a dramatic rise in internet adoption between 1995 and 2000, the dot-com crash of 2000 saw $1.7 tn lost in financial markets (McCullough, 2018). Of the 7000–10,000 online enterprises that had been launched in the late 1990s, by mid-2003, approximately 4,800 of them had either been sold or collapsed (ibid.).

In the context of sweeping digital disruption and two decades of accelerating financialisation, a more nuanced picture of a changing music economy emerges. It is a picture that indicates the febrile nature of the digital landscape in the early 21st century, where financial imperatives, such as rapid expansion and economic intensification, were creating an increasingly unstable economic environment for music, and where a recalibration of commerce and rights management began to emerge. Bringing this evolution up to the present day, RIAA analysis shows that revenue from synchronisation grew steadily between 2009 and 2022 (RIAA, 2023), and in Music Business Worldwide, reported that the US Music Publishing industry had doubled in size, from $2.2 bn in 2013 to $4.7 bn 2021 (Ingham, 2022b). Although the figures above demonstrate that streaming has come to dominate the music economy, its growth is not unique, and as we shall see, it is the ongoing evolution of multiple revenue streams that is one of the key challenges for the streaming economy.

Beyond these economic developments, it is also important to recognise the role that internet piracy had to play in the transition towards our current platform environment. In 1998, the RIAA had started suing online companies for hosting free music. However, against the dominant narrative that Napster and other peer-to-peer networks such as Aimster, AudioGalaxy, Morpheus, Grokster, iMesh, Kazaa, LimeWire and Scour were largely the cause of a significant haemorrhaging of value in the music industry in the late '90s / early '00s, Turner's view is that – as evidenced by the increased appetite for acquisitions which had coloured the previous decade – the major record labels were already struggling financially (Turner, 2022a). For Turner, Napster was less an existential threat than it was a confirmation of decline and an indication of the future direction of the global music industry. Indeed, he notes that when Napster was sued in December 1999, it had 50,000 users, but by the time it peaked in Feb 2001, it had 26.4 mn users (ibid.). Furthermore, he cites early 21st-century record sales in the UK as evidence that Napster was not the cause of the rapid decline in physical sales. Instead, he attributes this decline to the emergence of the iTunes Store:

> The drop in record sales in the UK does not coincide with the rise of Napster. It happens in the mid-2000s with the arrival of the iTunes Store, and the mass adoption of iTunes.
>
> *(Turner, 2022b)*

It is also significant that Daniel Ek was directly influenced by his experience of using Napster as a teenager. His ambition for Spotify was to harness what he saw as Napster's ability to connect music fans with what felt like an almost boundless vault of music content, while at the same time, ensuring that artists' rights and capacity to monetise their content was protected:

> I want to replicate my first experience with piracy. What eventually killed it was that it didn't work for the people participating with the content.
>
> *(Ek, in Van der Sar, 2018)*

Along with Napster, The Pirate Bay, a Swedish file-sharing website founded in 2003 that enabled users to participate and benefit from peer-to-peer sharing using Bit Torrent, played an important part in Spotify's early development. Following a raid by Swedish police on The Pirate Bay in 2006, which resulted in the site going offline for three days, a pro-piracy sentiment began to develop in Sweden, to such an extent that five out of its seven political parties expressed an interest in reviewing the country's copyright laws (Ekman, 2006). In the face of such uncertainty regarding the future sanctity of artists' rights, Per Sundin, CEO of Sony in 2006, suggested that the major record labels saw Spotify as a way to recoup losses, and were therefore happy to support the new startup by enabling it to licence their artists for its platform:

> If Pirate Bay had not existed or made such a mess in the market, I don't think Spotify would have seen the light of the day. You wouldn't get the licenses you wanted.
>
> *(Sundin, in Van der Sar, 2018)*

As this brief history shows, digital streaming developed in response to a range of factors. Beyond the conventional narrative that digital piracy – particularly Napster – 'killed' the music industry in the early 21st century (Forde, 2019), increased financialisation, combined with the fall-out from the dot-com boom and bust of the early 2000s, clearly had a significant impact on the music ecology of commerce, rights and user experience. While Spotify responded to the threat posed to artists' rights by the wave of digital piracy that swept through the industry in the early 2000s, its design was also influenced by the pirates, in terms of ease of access, and of its capacity to channel and reflect fans' appetites for discovering, sharing, categorising and compartmentalising music.

4.   A Brief Note on Spotify

Perhaps more than any other streaming platform, Spotify is most frequently associated with the wave of change and disruption that has swept through the music industry during the last two decades. In its relatively brief lifespan, Spotify's music on-demand ethos has contributed to giving music greater ubiquity in human life than at any other point in our history by embedding playlist culture as a global phenomenon, and continually engaging listeners and music fans with a range of user-centred listening experiences and innovative features. For example, the platform's 'Spotify Wrapped' campaign, which has run annually since 2016, provides Spotify users with a range of insights into their listening habits across the year. With 30 mn users accessing Spotify Wrapped in 2017, engagement with the campaign has consistently grown, reaching 156 mn users as of 2022 (De Guzman, 2023).

However, the platform's approach to creating new opportunities for listening, and innovating to build new audiences and markets has also divided the music community. In 2020, the UK government enquiry into streaming royalties saw Tom Gray's #BrokenRecord Campaign trigger a UK government Select Committee inquiry into the Economics of Streaming, characterised by in-depth scrutiny into payment models. Although the focus was on the music streaming sector more broadly, often, Spotify served as a point of focus. For example, the following passage from the concluding statement in the '#BrokenRecord Campaign response to CMA scoping document: Music and the Streaming Market Study', suggests that, to a large extent, Spotify – through its relations with major music labels – has set pricing and revenue rates that have become benchmarks for the wider streaming sector:

> Overall, we think it is clear that the dominant power of major music groups, the pricing of services and allocation of income are the most important issues facing music SMEs in subscription streaming. Spotify's unchanged price in over a decade [. . .] has produced around 25% deflation and this is before you consider their near halving of Average Revenue Per User which leads to the much-publicised collapse in 'per-stream rates'.
>
> *(#BrokenRecord Campaign, 2021)*

Similarly, when Daniel Ek suggested in 2020 that music had entered a new era of production, in which musicians and producers will no longer put out an album every three to four years but will instead engage with the attention internet's always-on workflow, his ideas drew considerable criticism. In response, the NME ran a highly critical comment piece of Ek's proposal, titled 'Spotify's Daniel Ek wants artists to pump out "content"? That's no way to make the next "OK Computer"' (Beaumont, 2020), while Musicradar.com, the online music magazine published a story that featured several high-profile musicians' Twitter / X responses (Laing, 2020).

These controversies are not new. And while these reactions testify to how, for some, Spotify has become the villainous face of everything that is wrong with the contemporary streaming landscape, from its beginnings, the platform can also be understood as a radical experiment in how to make a new kind of music industry; a platform which – unlike its peer-to-peer sharing predecessors – paid artists, locked down its digital content and its artists' IP, and inaugurated a new era of listener engagement and consumer behaviour.

5. The Streaming Curve: Current Challenges

As of 2023, it is clear that Spotify and its DSP relatives are no longer the disruptors they once were. With the ceaseless pressure to generate profit, to increase their payments to artists, alongside the growth in non-DSP markets for music – in particular the likes of TikTok – and the rapid evolution and pervasive spread of AI into the digital music ecosystem, streaming platforms in the form we have come to know them, the very likely now-familiar experience of music streaming itself, stand to be disrupted by new kinds of insurgent start-ups, or by a new industry paradigm. At what point does the disruptor become the dinosaur?

If the rapid evolutions in music – from the selling of physical products, to digital piracy and online streaming– which characterised the first decade of the 21st century are the digital music industry's Epoch A, then the streaming environment itself now faces a range of challenges that are slowing growth. For while the likes of Spotify, Amazon Music, Apple Music, Deezer and Tidal have not stopped modifying the streaming platform environment in their quest to build the optimum digital music model, the fundamental DSP model is facing challenges on various fronts; challenges that may well catalyse a transition from music streaming's own Epoch A to Epoch B. At the time of writing, this inflection point can be characterised by the following challenges:

6. Subscriptions

In June 2023, in its financial statement for Q2 2023, Spotify confirmed that it was operating in 184 countries and territories, and that its platform had 551 mn monthly active users ('MAUs'), partly comprised of 220 mn Premium Subscribers (Spotify, 2023b). This was a 2.2% increase on its Q1 figure of 210 mn Premium Subscribers (Stassen, 2023a), a growth rate that contributed to a 17% year-on-year increase as of the Q2 figures (Spotify, 2023b).

Spotify's continued growth has enabled it to remain ahead of its competitors, who, as of February 2023, reported the following numbers of US subscribers:

- Spotify 44.4 mn
- Apple Music 32.6 mn
- Amazon Music 29.3 mn
- YouTube 8.5 mn
- Pandora 2.4 mn
  (Stassen, 2023a)

Given this lead, it is therefore notable that both Apple and Amazon raised their prices in US and UK markets in 2023 – from $9.99 to $10.99 and from £9.99 to £10.99 respectively – and both DSPs continued to report post-increase growth (ibid.). By late July 2023, Spotify had followed suit, increasing its Premium Subscription tier in 52 global markets, including the US, the UK, Canada, Brazil and several European and Asian territories (Stassen, 2023b). Spotify's price rises follow calls throughout 2023 for the streaming sector to both increase its subscription prices (Music Business Worldwide, 2023), and to explore alternative approaches to pricing. For example, a tiered structure that would acknowledge the difference in value between an Ed Sheeran track and a calming recording of nature sounds (Ingham, 2023b). It is clear that, due to its position as market leader, much of the debate focuses on Spotify. While its lack of leadership in introducing higher prices suggests a degree of nervousness at the platform as regards maintaining its market dominance at the expense of generating extra revenue, at the same time, owing to global inflation rates, commentators remain unconvinced that the recent increases will lead to positive benefits either in terms of the company's own profits or to increases in income for rightsholders (Forde, 2023).

Beyond these ongoing challenges in the Premium Subscription landscape, David Turner has reported that ad-supported streaming is facing difficulties of its own. Confirming that Gaana, an Indian DSP, Deezer and Resso, who along with TikTok, which is part of ByteDance, have all removed their ad-supported free tiers (Turner, 2023b). Clearly, the dilemma for DSPs is that revenue from subscriptions is insufficient to deliver meaningful and sustained improvements to their finances, but at the same time – as Spotify's hesitancy demonstrates – increasing prices might jeopardise their subscriber figures. At the same time, as Turner suggests, ad-supported streaming is far from being a reliable source of income, where DSPs continue to struggle to support free tier into premium tier subscribers and where the costs of ad-supported markets can often outweigh the gains (ibid.).

7. Overload

In May 2023, the entertainment data and insights company, Luminate, reported that over 120,000 tracks were being uploaded to streaming services per day (Fu, 2023). This represented a 28.5% increase on 2022 figures, where the daily upload average was 93,400 and overall, meant that, in the first three months of 2023, over 10 mn tracks had been uploaded to DSPs, including Spotify, Apple Music and YouTube Music (ibid.). Furthermore, BandLab Technologies, the company behind the social music creation app BandLab, reported that across two financing rounds in 2022 and 2023 it had raised $90 mn, taking the value of the company to $425 mn (Dalugdug, 2023b). With global user figures in excess of 60 mn users, and a corporate focus on 'mass music creation [and a] dedication to artists, aspiring musicians, and creators worldwide' (Kuok, in Dalugdug, 2023b), it would seem that BandLab's social music paradigm, which connects easy-to-access music production technology with easy-to-access audience engagement strategies, looks set to continue facilitating the entry of a wealth of amateur and semi-professional producers into the global streaming ecosystem.

While it is encouraging that so many artists and rightsholders are engaging with the opportunities to connect with audiences that DSPs' relatively low entry barriers afford, these figures do suggest that the streaming ecosystem is facing supply-side challenges. In other words, who is listening to all of this content? When we consider that, as of late 2022, almost 80% of this content had less than 50 monthly listeners and that 90% had fewer than 400 monthly listeners (Kennelty, 2022), the challenge of how to navigate this content-heavy landscape becomes even more stark – either from the perspective of the emerging artist seeking to develop their profile, or the music fan looking for new sounds. Indeed, even established rightsholders are faced with the complexities of ensuring that their content remains visible and audible in an increasingly saturated environment.

8. Artificial Music

And this over-saturation is not only the result of human endeavour.

In May 2023, Spotify took down two tracks that had been created by the AI music production app Boomy (Tencer, 2023). The tracks were removed from Boomy's 54-track 'This Is Boomy' playlist, and while the two tracks in themselves represent a tiny fraction of the music available on the platform, they speak of an increasingly pervasive problem facing DSPs. As of August 2023, since its launch in 2019, Boomy has created over 17 mn songs (Boomy, 2023) – reportedly over 13% of global recorded music (Tencer, 2023) – an indication of the speed at which AI services are challenging conventional music industry practices.

According to Music Business Worldwide, the Boomy takedown is indicative of two key challenges that AI services are bringing to streaming services: artificially produced music and artificially produced streams (Ingham, 2023a). Where artificially produced music stands to compete with human artists and traditional rightsholders for listener revenues, artificial, or fake streams created using bots have already been identified as a means by which revenue is created fraudulently for low-quality or indeed AI-generated content (Dredge, 2023 and Ingham, 2023a). In his 2023 New Year update to staff, Sir Lucian Grainge, Chairman and CEO of Universal Music Group, said that the challenge posed to the music industry by artificial intelligence essentially centres on a divide between 'those committed to investing in artists and artist development versus those committed to gaming the system through quantity over quality' (Grainge, in Rys, 2023), a sentiment that emphasises the problematic way in which AI systems are both overloading and gaming current streaming protocols. In this context, it is worth noting two further recent developments in music AI. In May 2023, Google's high-fidelity music generator, MusicLM became available for public use (Yim and Manickavasagam, 2023). The tool uses text prompts to create music, and having piloted it with established musicians, the production team launched MusicLM into the public realm to garner user response which could be fed back into the project's development process (ibid.). As noted above with the growth of BandLab, the MusicLM project demonstrates that, for the time being, disruptions in streaming caused by AI look set to proliferate in the near future, and will continue to force DSPs to rethink and recalibrate their operating models in order to meet and combat the challenges to their supply and stream-count protocols.

9. Payments

In 2020, the UK Parliament conducted an inquiry into how streaming services compensated rightsholders, releasing a report in 2021 entitled 'The Economics of Music Streaming' (UK Parliament, 2021). For the musicians involved in the inquiry, it was an opportunity to draw attention to, and problematise, the intricacies of the models being used by DSPs to pay artist royalties. As of 2023, debates around artist and rightsholder payments have become much more commonplace, and a set of terms that describe the various models used by DSPs are now familiar to both industry professionals and music fans alike. The 'pro-rata' model – which Spotify refers to as 'Streamshare' (Spotify, 2023c) – describes how the total income generated in a given period (from subscriptions and advertising) is aggregated and then shared with rightsholders in proportion to the total number of streams their tracks receive during that period. The user-centric model – which SoundCloud refers to as 'Fan Powered Royalties' or 'FPR' (Thakrar, Cirisano and Gresham, 2022) – directly connects a user's subscription fee to the tracks they listen to, thereby incentivising artists to build their community and increase the number of streams their tracks are receiving. At the time of writing, SoundCloud has implemented an FPR system, and is working with Warner Music Group whose artists – including Dua Lipa, Madonna and Ed Sheeran – are paid via FPR when streamed on SoundCloud. According to a report on Fan Powered Royalties produced by MIDiA in 2022, the FPR model largely benefits artists whose music and profile attracts 'superfans'; that is, artists who have between 1000 and 10,000 active listeners, although the report does confirm that artists whose fanbase ranges from 100 – 100,000 did see an overall increase in royalty payments from an FPR system (ibid.). In May 2023, the digital music licensing agency, Merlin, signed a partnership agreement with SoundCloud, meaning that its membership, which includes Cinq Music, Amuse and Foundation Media, will also benefit from the FPR method (Dalugdug, 2023a).

Beyond questions of creating a more balanced payment system, FPR is also seen as having benefits in the battle against streaming fraud (Lavoie, 2023). Where pro-rata is used by the likes of Spotify and Apple Music, bots generating fake streams are able to exploit the pooled store of income, and thereby syphon-off revenue that should otherwise be going to genuine rightsholders. Again, by creating a direct connection between a track's stream count and an artist's royalty payments, FPR can function as a deterrent, simultaneously making it harder to generate income from fake streams – since FPR rewards fan engagement – and simultaneously easier to identify fraudulent activity.

SoundCloud's success in deploying a user-centric approach is clearly having a broader influence, and throughout 2023, Lucian Grainge has continued to voice UMG's commitment to developing what he refers to as an 'artist-centric' model. In July Grainge listed three priorities that would underpin this model – implementing a fair payment system that connects 'real artists with real fanbases', clamping down on fraudulent activity, including what he calls 'bad actors' who upload 'noise and other content that has no meaningful engagement from music fans', and improving user experience so as to increase discovery, support 'real' artists – and went on to confirm that Spotify was now formally partnering with UMG on the latter's artist-centric initiative (Grainge, in Hits Daily Double, 2023).

While this growing interest in user-centric methods of payment appears to be the positive response to artist remuneration demanded by the #BrokenRecord campaign (which catalysed the 2020 UK Parliament inquiry), the model is not without its own faultlines. Where MIDiA's 2022 report showed that 56% of artists benefitted from FPR as opposed pro rata, as Tim Ingham points out, this means that 44% were worse off. While MIDiA's 100–100,000 bell curve of optimum fan engagement means that the more artists can do to cultivate active fans and listeners, the more they will benefit, for major artists, whose fanbase may well exceed the 100,00 upper limit of engaged and committed fans, it may be possible that they stand to make a loss (Ingham, 2022a). Furthermore, David Turner, writing in 2019, reported that three separate studies into the impact of user-centric streaming on the top 5000 artists on streaming platforms in Norway, Denmark and Finland all showed that, while earnings for the middle-tier of artists increased and earnings for the top tier varied between marginal increase and marginal decline, earnings for the bottom tier of artists, and those outside of the top 5000, all suffered a loss of income as a result of increases to the top two tiers (Turner, 2019a). Clearly, the ultimately equitable payment model for streaming services remains some way off.

10.   Competition

Beyond challenges that are, to a greater extent, native to the streaming paradigm – matters of supply, accounting and payment protocols – increasingly, streaming services are also having to navigate external pressures, as social media platforms disrupt much more established patterns of music consumption. In the first two financial quarters of 2023, the meteoric growth in streaming revenues that was in evidence in 2021, which was largely due to a post-pandemic surge in economic growth, along with upfront payments for what MIDiA refers to as 'non-DSP streaming' – in other words, the deals between rights holders and the likes of TikTok, Meta and Snap – has slowed. At the start of the year, MIDiA compared growth in streaming revenues for Q3 2021 (31%) with Q3 2022 (7%), demonstrating, that while we are not yet in a period of decline as such, 2023 would be a challenging year for DSPs and the streaming environment (Mulligan, 2023). As Mulligan had suggested in October 2022, a key issue that continues to face DSPs is the use of music in non-DSP environments in ways that – in DSP terms – are still relatively non-standard (for example, duets, lip-syncs and components and 'snippets of songs') (Mulligan, 2022). Whether or not streaming services are able to respond to artists' calls for fair remuneration, rise to the challenge of fake artists and fake streams, settle on a price point that works for everyone and somehow manage to deal with the growing overload of content, it seems that the more fundamental challenge may simply be that music consumers are increasingly choosing to consume music in ways that are not native to the streaming environment. Sensing that this fragmentation only looks set to continue, if not grow, Mulligan's solution was to suggest that 'the music industry needs a new format' (ibid.).

A search for 'TikTokification' on Google is rewarded with pages of results, not least the following definition from the Urban Dictionary:

> The phenomenon in which social media platforms gradually become more like TikTok in order to become more relevant, even though no one wanted them to.
>
> *(GenWaller, 2022)*

As 2023 has progressed, it has become apparent that Spotify is determined to adapt to TikTok's paradigm-shifting approach to designing users' experience of content within a social context. In March, this led to Spotify 'TikTokifying' its Home page, introducing vertical feeds for music and podcasts and making use of AI to deliver personalised recommendations based on users' listening habits and preferences (Singh, 2023). In a move to shore-up its market-leading position with young listeners, the platform also announced a new video podcast feature aimed at younger users, and reported that it was working with 70,000 active video creators (Shapiro, 2023). In June, Spotify was also reported to be considering hosting full-length music videos as part of its expanded media offering, enabling it to compete more directly with YouTube (Carman, 2023), a further indication that the music streaming ecosystem, which once might have satisfied users by providing them with music as a self-contained experience, may now need to accommodate a variety of adaptations in order to survive.

Writing on the MIDiA blog, Tatiana Cirisano engages with the likes of TikTok and YouTube to consider streaming providers' pricing mechanisms. She asks whether TikTok's Creator Fund, which in its revamped form only starts rewarding creators once their follower numbers exceed 100,000, could be a model for streaming services (Cirisano, 2023). Similarly, she remarks that YouTube's ad-revenue sharing programme is a means to incentivise creators by generating financial benefit for both creator and YouTube alike (ibid.). However, while Cirisano acknowledges that creator funds are time-limited and not permanent revenue solutions for artists and creators, in the context of Robert Kyncl's comments on the difference in value between an Ed Sheeran song and a recording of rain falling on a roof, might such a difference be another way of framing – and engaging with – the value challenge? In other words, would it benefit streaming services to consider limiting payouts to artists who have met a certain threshold of followers? And should there be other payment-related thresholds, for example, relating to quality and song length? As Mulligan suggested, would a new music format designed to accommodate the way that TikTok and Instagram users combine video with 15–30 second song snippets be another solution (Mulligan, 2022)?

With creator culture showing no signs of slowing down, in order to combat, and – in the words of the Urban Dictionary – 'become more relevant', some kind of format shift that accounts for user modification and customisation, accompanied by a continued focus on diversification of streaming's core offering will no doubt be necessary.

11. Conclusion

Using Salk's S curve to look at how streaming has developed since the early 2000s enables us to plot a development trajectory that has moved from rapid growth and mass adoption to a period in which the now familiar and conventional streaming paradigm is being challenged on a variety of fronts. I have written elsewhere on the concept of creative destruction, and how processes of technological change are frequently characterised by repeating patterns of disruption (Lovett, 2019a, 2019b) – where a disrupting technology is itself disrupted by an insurgent technology (for example, the portable disc drives that stole market share from IBM mainframe systems in the 1980s that have now been largely ousted by the global turn to cloud computing). In this sense, we can map the development of streaming to this familiar pattern of creative destruction. As the disruptor that finally unbalanced the music industry's reliance on physical goods, streaming in its current form as the dominant and incumbent force in music, does indeed look set to be replaced – or at least radically modified – by a new paradigm that will satisfy both audiences, artists and rightsholders alike. As such, where the likes of Spotify might once have worn their disruptor status proudly, challenging the music industry to move beyond physical sales, and into a new walled ecosystem of pay-per music consumption, it seems now that the streaming curve is indeed slowing down and starting to flatten out.

Beyond the idea that creative destruction and disruption occur in ever-repeating cycles, the S curve enables us to think about how change happens, to anticipate periods of declining growth, moments of turbulent transition and potentially more settled long-term futures. As we have seen in 2023, the challenges for a streaming-first company like Spotify are mounting. Unlike Apple, Amazon and Alphabet, Spotify's business has been entirely built, and continues to be focused, on streaming. However, it is yet to break even, let alone make a profit, reporting a $256 mn loss in 2022 (Iqbal, 2023) and an operating loss of $169 mn in Q1 2023 (Spangler, 2023). For a company so completely reliant on streaming, this remains problematic. Indeed, as far back as 2019, David Turner was reflecting:

> If a company that was used by hundreds of millions of people across the globe couldn't quite turn a profit, then how exactly is it competing with companies like Apple, Amazon, and Google? If Spotify were to become the largest, and perhaps even the sole, music streaming platform in the world, that wouldn't even begin to poke at the core businesses of their competitors.
>
> *(Turner, 2019b)*

While Spotify's innovative approach has given rise to some of the most fan-centred music experiences across the history of streaming, thinking about its development in terms of an S curve suggests that this period of turbulence may well increase and precipitate radical change. Streaming, as a core commercial focus for a company like Spotify, may ultimately founder before it ever really becomes established as a fully-functioning model for digital music consumption. As such, it may be that the streaming curve is in terminal decline, and that ultimately, streaming will not be able to adapt and change; it will simply have to transform into something else in order to find a sustainable and long-term future that facilitates digital music consumption on a global scale. MIDiA's proposal for a new format is certainly a compelling argument designed to embrace the abundance of musical experiences that a social-first audience increasingly demands, while Spotify's TikTokification of its interface also acknowledges the changing trends in in-app user experience. However the future of streaming plays out, Jonas Salk's S curve is a valuable tool that enables us to recognise patterns of change. The streaming curve, characterised by rapid growth in the early 2000s, followed by a slowing-down of adoption and widespread turbulence through the late 2010s and into the 2020s now suggests that a more paradigmatic transformation is likely, one that could yet precipitate a more settled future for digital music.

At this point, it appears that the key challenge for streaming is abundance; an abundance of music being uploaded by human or non-human creators, and of artists wanting to be paid more equitably for their creative labours, an abundance of fraudulent activity either uploading fake music, or using bots to generate fake stream counts, and an abundance of competing paradigms for music consumption, particularly from the likes of TikTok. For the likes of Spotify, Deezer, Tidal and other streaming-first platforms, it seems that building a long-term future for themselves will involve engaging with and solving the challenge that this super-abundance – an abundance of abundance – brings. Creating a multi-stakeholder model that satisfies consumers who want to use music in a variety of ways as well as artists who want to create an income from music however large their audience, is a key challenge, alongside staying relevant and maintaining market share in an environment where technology is pushing evermore towards a social-first paradigm. Essentially, to secure a sustainable future for themselves, streaming services need to provide everything for everybody, all the time. The challenge is significant, and it remains to be seen whether a streaming-first incumbent has the necessary flexibility to adapt to the new digital music ecology, or whether a social-first disruptor, whose approach to content management is user, rather than content-led, will lead music into a new era of similarly flexible music consumption.

References

Beaumont, M. 2020. 'Spotify's Daniel Ek wants artists to pump out "content"? That's no way to make the next "OK Computer"'. Available at www.nme.com/features/spotify-daniel-ek-three-to-four-years-controversy-criticism-2721051 (accessed August 2023).

Boomy Corporation. 2023. 'Boomy'. Available at https://boomy.com (accessed August 2023).

#BrokenRecord Campaign. 2021. '#BrokenRecord Campaign response to CMA scoping document: Music and the Streaming Market Study'. Available at https://assets.publishing.service.gov.uk/media/625419968 fa8f54a922789a6/Broken_Record_Campaign.pdf (accessed August 2023).

Carman, A. 2023. 'Spotify is in talks to test full-length music videos in app'. Available at www.bloomberg. com/news/articles/2023-06-30/spotify-is-in-talks-to-test-full-length-music-videos-in-app?sref=10INA hZ9&in_source=embedded-checkout-banner (accessed August 2023).

Cirisano, T. 2023. 'Should streaming take a cue from TikTok's revamped creator program?' Available at www. midiaresearch.com/blog/should-streaming-take-a-cue-from-tiktoks-revamped-creator-program (accessed August 2023).

Cirisano, T., Das, S., Gresham, P., Griffin, S., Kahlert, H., Langston, A., Millar, A., Mulligan, M., Mulligan, T., Rothwell, K., Thakrar, K., Severin, K., and Woods, B. 2023. '2023 MIDiA predictions: Pivot point'. Available at https://midiaresearch.com/reports/2023-midia-predictions-pivot-point (accessed August 2023).

Dalugdug, M. 2023a. 'SoundCloud, Merlin strike global deal to bring fan-powered royalties to indie labels'. Available at www.musicbusinessworldwide.com/soundcloud-merlin-strike-global-licensing-deal-to-bring-fan-powered-royalties-model-to-indie-labels/ (accessed August 2023).

Dalugdug, M. 2023b. 'BandLab raises another $25m, valuing music creation company at $425m'. Available at www.musicbusinessworldwide.com/bandlab-raises-another-25m-valuing-music-creation-company-at-425m/ (accessed August 2023).

De Guzman, C. 2023. 'Your complete guide to Spotify wrapped, 2023'. Available at https://time.com/6340 656/spotify-wrapped-guide-2023/ (accessed January 2024).

Dredge, S. 2023. 'Streaming subscriptions could be 'affordable luxuries' in 2023'. Available at https:// musically.com/2023/01/11/streaming-subscriptions-could-be-affordable-luxuries-in-2023/ (accessed August 2023).

Ekman, I. 2006. 'Politicians smell votes in Sweden's file-sharing debate – Technology – International Herald Tribune'. Available at www.nytimes.com/2006/06/18/technology/18iht-levies.1994335.html (accessed August 2023).

Forde, E. 2019. 'Oversharing: How Napster nearly killed the music industry'. Available at www.theguardian. com/music/2019/may/31/napster-twenty-years-music-revolution (accessed August 2023).

Forde, E. 2023. 'What will Spotify's price rise mean for its recording artists and songwriters?' Available at www.theguardian.com/music/2023/jul/28/what-will-spotify-price-rise-mean-for-its-recording-artists-and-songwriters (accessed August 2023).

Fu, E. 2023. '120,000 new tracks are being uploaded to streaming services per day'. Available at https://cons equence.net/2023/05/music-streaming-services-uploads/ (accessed August 2023).

GenWaller. 2022. 'Tiktokification'. Available at www.urbandictionary.com/define.php?term=Tiktokification (accessed August 2023).

Hits Daily Double. 2023. 'Grainge on UMG's "Artist-Centric" approach'. Available at https://hitsdailydou ble.com/news&id=336930&title=GRAINGE-ON-UMG'S-"ARTIST-CENTRIC"-APPROACH?ref=pen nyfractions.ghost.io (accessed August 2023).

IFPI. 2023. *Global Music Report 2023*. Available at www.ifpi.org/resources/ (accessed August 2023).

Ingham, T. 2022a. 'On SoundCloud's new deal with Warner Music (or why the debate over fan-powered royalties is more complicated than "fairness")'. Available at www.musicbusinessworldwide.com/on-soun dclouds-new-deal-with-warner-music/ (accessed August 2023).

Ingham, T. 2022b. 'The US music publishing industry generated $4.7bn last year – but the record industry grew twice as fast'. Available at www.musicbusinessworldwide.com/the-us-music-publishing-industry-generated-4-7bn-last-year-but-the-record-industry-grew-twice-as-fast/ (accessed August 2023).

Ingham, T. 2023a. 'The Boomy/Spotify streaming fraud debacle proves 'pro-rata' must go – urgently'. Available at www.musicbusinessworldwide.com/the-boomy-spotify-streaming-fraud-debacle-proves-pro-rata-must-go-urgently/ (accessed August 2023).

Ingham, T. 2023b. '"An Ed Sheeran stream is not worth the same as a stream of rain falling on a roof": Robert Kyncl says music streaming payout and pricing models must, and will, change'. Available at www.mus icbusinessworldwide.com/ed-sheeran-stream-is-not-worth-the-same-as-a-stream-robert-kyncl-says-music-streaming-payout/ (accessed August 2023).

Iqbal, M. 2023. 'Spotify revenue and usage statistics (2023)'. Available at www.businessofapps.com/data/spotify-statistics/ (accessed August 2023).

Kennelty, G. 2022. 'Statistically, nobody is listening to most bands on Spotify'. Available at https://metalinjection.net/its-just-business/statistically-nobody-is-listening-to-most-bands-on-spotify#:~:text=And%2C%20in%20fact%2C%2090%25,1%20millionth%20of%20the%20platform. (accessed August 2023).

Laing, R. 2020. 'Spotify CEO: "You can't record music every three or four years and think that's going to be enough"'. Available at www.musicradar.com/news/spotify-ceo-you-cant-record-music-every-three-or-four-years-and-think-thats-going-to-be-enough (accessed August 2023).

Lavoie, A. 2023. 'Why fake streams hurt every developing artist'. Available at https://blog.landr.com/fake-streams/ (accessed August 2023).

Lovett, M. 2019a. 'Disruptive Blockworks: Blockchains and networks/acceleration and collision', in Ragnedda, M. and Destefanis, G. (eds). *Blockchain and Web 3.0: Social, Economic, and Technological Challenges*. Routledge.

Lovett, M. 2019b. 'Disruption as Contingency: Music, Blockchain, Wtf?', in Hepworth-Sawyer, R., Hodgson, J., Paterson, J. and Toulson, R. (eds). *Innovation In Music: Performance, Production, Technology and Business*. Routledge.

McCullough, B. 2018. 'A revealing look at the dot-com bubble of 2000 – and how it shapes our lives today'. Available at https://ideas.ted.com/an-eye-opening-look-at-the-dot-com-bubble-of-2000-and-how-it-shapes-our-lives-today/ (accessed August 2023).

Meier, G. 2022. 'Sound money: On Bandcamp, Neil Young derivatives and the financial imagination of music production' Available at https://bellonamag.com/sound-money?ref=pennyfractions.ghost.io (accessed August 2023).

Mulligan, M. 2022. 'The music industry needs a new format'. Available at https://musicindustryblog.wordpress.com/2022/10/20/the-music-industry-needs-a-new-format/ (accessed August 2023).

Mulligan, M. 2023. 'Has the streaming slowdown arrived?' Available at https://midiaresearch.com/blog/has-the-streaming-slowdown-arrived (accessed August 2023).

Music Business Worldwide. 2023. 'These worrying stats send a clear message: the music biz needs a Spotify price rise . . . now'. Available at www.musicbusinessworldwide.com/podcast/these-worrying-stats-send-a-clear-message-the-music-biz-needs-a-spotify-price-rise-now/ (accessed August 2023).

RIAA. 2023. 'U.S. Music Revenue Database'. Available at www.riaa.com/u-s-sales-database/ (accessed August 2023).

Rys, D. 2023. 'Lucian Grainge calls for 'updated model' for music industry: Read his memo to UMG staff'. Available at www.billboard.com/pro/lucian-grainge-umg-full-staff-memo-2023-read-message/ (accessed August 2023).

Salk, J. 2019. *Planetary Health: A New Reality. Challenges*, [online] 10(1), p.7. https://doi.org/10.3390/challe10010007.

Salk, J. and Salk, J. 1981. *World Population and Human Values: A New Reality*. HarperCollins.

Shapiro, A. 2023. 'Spotify is going big on video podcasts'. Available at www.theverge.com/2023/3/8/23630901/spotify-markiplier-video-anchor-julia-fox-batman-homepage-tiktok (accessed August 2023).

Singh, S. 2023. 'Spotify announces new TikTok-like vertical interface'. Available at www.nme.com/news/music/spotify-new-vertical-feed-interface-tiktok-3410470 (accessed August 2023).

Spangler, T. 2023. 'Spotify tops Q1 subscriber targets to hit 210 million paying users, revenue misses'. Available at https://variety.com/2023/digital/news/spotify-q1-2023-earnings-1235593783/ (accessed August 2023).

Spotify. 2023a. *Financial Statements: Q1*. Available at https://investors.spotify.com/financials/default.aspx#quarterly-results (accessed August 2023).

Spotify. 2023b. *Financial Statements: Q2*. Available at https://investors.spotify.com/financials/default.aspx#quarterly-results (accessed August 2023).

Spotify. 2023c. 'Royalties'. Available at https://support.spotify.com/us/artists/article/royalties/ (accessed August 2023).

Stassen, M. 2023a. 'Spotify had 44.4m US subscribers in February, Apple Music had 32.6m, according to new data', Available at www.musicbusinessworldwide.com/spotify-had-44-4m-us-subscribers-in-february-apple-music-had-32-6m/ (accessed August 2023).

Stassen, M. 2023b. 'Spotify officially raises Premium Subscription prices in the US, and in several markets across South America, Europe and Asia'. Available at www.musicbusinessworldwide.com/spotify-off icially-raises-premium-prices-in-the-us-and-in-several-markets-across-south-america-europe-and-asia1/ (accessed August 2023).

Tencer, D. 2023. 'AI music app Boomy has created 14.4m tracks to date. Spotify just deleted a bunch of its uploads after detecting "stream manipulation"'. Available at www.musicbusinessworldwide.com/ai-music-app-boomy-spotify-stream-manipulation/ (accessed August 2023).

Thakrar, K, Cirisano, T and Gresham, P. 2022. *Building a fan economy with Fan-Powered Royalties.* Available at www.midiaresearch.com/reports/building-a-fan-economy-with-fan-powered-royalties (Accessed August 2023).

Turner, D. 2019a. 'The false promise of user-centric streaming'. Available at https://pennyfractions.ghost.io/the-false-promise-of-user-centric-streaming/ (accessed August 2023).

Turner, D. 2019b. 'A correct history of music streaming'. Available at https://pennyfractions.ghost.io/a-corr ect-history-of-music-streaming/ (accessed August 2023).

Turner, D. 2022a. 'Recession looms over the music industry (Part 1)'. Available at https://pennyfractions.ghost.io/music-industry-recession-part-1/ (accessed August 2023).

Turner, D. 2022b. 'Money for nothing: Web 3.Bro with David Turner. [Podcast]. February 2022'. Available at https://open.spotify.com/episode/50iQKCq5bA5lDaBdCIuwyS?si=d4f91975e54b4f9d (accessed August 2023).

Turner, D. 2023a. 'The developing "crisis" of music streaming'. Available at https://pennyfractions.ghost.io/the-developing-crisis-of-music-streaming/?ref=penny-fractions-newsletter (accessed August 2023).

Turner, D. 2023b. 'Ad-supported music streaming is broken'. Available at www.musicbusinessworldwide.com/ad-supported-music-streaming-is-broken/ (accessed August 2023).

UK Parliament. 2021. 'Economics of music streaming'. Available at https://committees.parliament.uk/event/15732/formal-meeting-oral-evidence-session/ (accessed August 2023).

United Nations. 2023. 'Global issues: Population'. Available at www.un.org/en/global-issues/populat ion#:~:text=Our%20growing%20population&text=The%20world%27s%20population%20is%20expec ted,billion%20in%20the%20mid%2D2080s. (accessed August 2023).

Van der Sar, E. 2018. 'How the Pirate Bay helped Spotify become a success'. Available at https://torrentfreak.com/how-the-pirate-bay-helped-spotify-become-a-success-180319/ (accessed August 2023).

Yim, K. and Manickavasagam, H. 2023. 'Turn ideas into music with MusicLM | Experiment today by describing a musical idea and hearing it come to life'. Available at https://blog.google/technology/ai/musi clm-google-ai-test-kitchen/ (accessed August 2023).

3

MISSED OR POSTPONED INNOVATION?

The Rise and Fall (and Rise?) of Live Streamed Music Events in Italy

Francesco D'Amato

1. Introduction

During the pandemic, the lockdown and the social distancing obligations made the normal carrying out of concerts impossible and – consequently – encouraged an exponential growth of live streaming of music gigs of extremely heterogeneous nature and dimensions, in some cases replicating and expanding pre-existing forms of online music performances, in others experimenting with new ways and models. This surge of live streamed concerts has understandably attracted the interest of several scholars, fueling the development and publication of research covering a range of topics, such as: the characteristics of online music performances (Rendell, 2021; Cireddu, 2023) and the strategies developed from musicians approaching the challenge of performing live for a distant audience (Kjus et al., 2022); the ways of audiences participation (Vandenberg, 2020; Rendell, 2021; Ferreira et al., 2022; Cireddu, 2023), their motivations and the values they attribute to this type of events (Perez-Monteagudo and Curras-Perez, 2022), their experiences of social connectedness (Vandenberg, 2020; Onderdijk et al., 2021); the effects of the platformization of live music activities on local and translocal music scenes (Mouillot, 2022).

A more neglected issue is the trajectory of the live streaming concert offer, from the pandemic to today. In Italy, as in many other countries, after the sudden exponential growth of such an offer, there was an equally drastic and sudden drop once the pandemic emergency was over, even though many professionals in the sector firmly believed that those types of content would have survived and continued to develop alongside normal concerts in physical proximity. As far as I know, there is a lack of accounts attempting to understand this trajectory by taking into consideration aspects relating to production and offering models. This is the topic addressed by this chapter.

What follows are the results of an exploratory investigation developed as a sort of spin-off of research concerning the participation in live streamed concerts of Italian musicians. That research, conceived in May 2020, should have included participant observation of online music events combined with questionnaires and interviews with participants of those same events, while the analysis of the supply side should have served only as part of the context. However, after a little over a year, the live streaming of concerts by established Italian musicians had suddenly ceased, despite the expectations of most of the music industry professionals, while many of the services that aimed to develop an innovative offer confronted numerous problems that arrested or slowed down their development. This led me to investigate the reasons for this failed or postponed innovation.

DOI: 10.4324/9781003396550-3

The analysis therefore refers to a specific local context and especially to the offer of the digital platforms which in that period (particularly from September 2020 to September 2021) invested the most in the streaming of concerts by popular Italian musicians, such as Italian A-Live and English Live Now. This part of the research consisted of eighteen interviews with different stakeholders, including platform managers interviewed several times in different periods, complemented with a wide review of articles from newspapers, industry journals, and blogs, documenting and commenting on the emerging phenomenon of the live streamed music events; however, in some points, I will also refer to the results of a questionnaire completed by 210 participants in live streamed concerts. The interviews were divided as follows: seven interviews with Italian managers of four different platforms involved in the live streaming of music events by Italian musicians, during the pandemic and in the immediate aftermath (A-Live, Live Now, Live All and Dice, which is mainly a ticketing company); five interviews with musicians involved in live streamed concerts or with their managers; three interviews with live promoters involved in the production of live streamed concerts; one interview with a staff member of a collecting society of authors and publishers; one interview with the head of the technological partner of the main Italian live streamed event (Heroes), and; one interview with the head of the Italian music venues association (KeepOn Live). The interviewees have been coded and referred to – in the cases of direct quotes – in order to preserve anonymity.

2. Innovations and Discourses Supporting the Rise of Live Streamed Concerts

Since the summer of 2020, there have been many live streamed concerts by established Italian musicians. In that context, A-Live and Live Now represented two examples of platforms that aimed to develop an innovative offer, providing new content and new ways to participate, which included – for example – the possibility of:

- seeing oneself and being seen by others within the performance space, through virtual backdrops;
- customizing the vision by selecting the shots;
- sound reactions – such as *clapping* and *ovation* – audible both by the artists and by other participants;
- interaction between participants via public and/or private live-chat, dedicated rooms, and the "watching together" feature;
- direct interaction with the musicians during meet and greet;
- enjoying performances characterized – for example – by massive use of augmented reality, aimed at altering the space and scenography of the performance.

The platforms offered new possibilities to musicians as well; for example, A-Live allowed them to create NFTs in real time during the performance, which could later be auctioned off. On the other hand, with regard to pricing models, the offer contemplated both access via pay-per-view, generally with differentiated prices according to the features of participation, and free access to events financed by sponsors.

The offer was contextualized by discourses which, on the one hand, supported the value of live streaming both for the audience and the musicians, while on the other formed the basis for the belief that participation in live streamed events would have complemented participation in events in physical proximity, once the pandemic had passed. The most recurring arguments were supporting the value of live streamed events for audience and musicians.

First, live streaming allows the participation of those who do not have the opportunity to attend in person, for various reasons and conditionings, such as distance, costs, various forms of disability, and work and family commitments. This consideration allowed platforms both to evoke the concepts of inclusiveness and democratization of access and to promise musicians new opportunities for expanding and diversifying their audience. Moreover, according to some observers, the particularly low price of live streamed concerts could have attracted people who may not be big fans of the musician but might be interested in learning more about them by attending their streamed performance. Regarding the first point, it is worth noting that 95% of respondents to our questionnaire confirmed that they had attended events extremely far from where they live, and in general this opportunity turned out to be by far the most appreciated benefit of online events (80%), while the most annoying aspects of offline events are mainly considered precisely "the time and difficulty necessary for travel" (80.4%) and "the cost of tickets" (64.9%). On the other hand, regarding the second point, just under a fifth of respondents claimed to have attended concerts by performers they knew little (13.6%) or not at all (5.8). Furthermore, the enjoyment of online events by little-known musicians could have been accentuated by the audience's willingness to search and try out new entertainment content during the lockdown periods.

Secondly, new monetization opportunities associated with digital events, such as digital merchandise and NFTs, would have further attracted musicians and their managers.

Third, the musicians would have been able to experiment with new ways of expression and of relating with the audience, while the audience would have had the opportunity to enjoy unusual concerts or to enjoy concerts in unusual ways, combining the comfort of participation from home with various opportunities for personalizing the vision and interacting with the performers and the other participants. Actually, the possibility of enjoying the events from the comfort of home turned out to be the second most appreciated feature of online events by the respondents to the questionnaire (52%), followed by "free or low-cost tickets" (37%) and "being able to see the performer well" (33%).

Finally, a quite recurrent 'argument' was: "since it worked for soccer, where many people still go to the stadium while others prefer to watch the game on television, why shouldn't it work for music"?

Indeed, many live streaming of Italian musicians have been considered successes. Just a few examples:

- in September 2020, the concert of the metal band Lacuna Coil sold a slightly higher number of tickets than it would have if sold out at the venue from which it was streamed, with a third of the audience connected from abroad;
- in the same month, the mega charity event Heroes, joined by almost forty musicians in support of music workers affected by the lockdown, sold nearly 40,000 tickets;
- in November of the same year, the concert of the popular band Negramaro aroused considerable hype for its use of augmented reality;
- in April 2021, the concert of the popular singer-songwriter Ultimo hit the record of tickets sold for a streaming event in Italy.

However, after just over a year, this trend was already completely exhausted.

3. So, What Went Wrong?

According to the managers of the live streaming platforms, the feedback from both the musicians involved and the participating audience had been positive. Our questionnaire confirmed a general appreciation by the public: referring to the last online music event attended, 63% of respondents considered themselves quite satisfied and 29% very satisfied, although 53% agree in stating that "the experience was pleasant but less intense than a live event attended in physical presence". Regarding these findings, it is worth noting that another research, comparing value perceptions in attending 'live' and 'online' music festivals, found that

> the perceived value affected satisfaction more in online festivals than in live festivals, perhaps because the participants' expectations about the value of online festival experiences were lower, so the same level of value translated into greater satisfaction with online festival.
>
> *(Perez-Monteagudo and Curras-Perez, 2022, p. 431)*

Whatever the case, despite the apparent satisfaction from both musicians and audiences, the main criticism, on which all the interviewees agree, concerned the *economic unsustainability* of the offer.

> The results were always decidedly negative, not even managing to cover the costs, which are still important for a good streaming event.
>
> *(C., live promoter and producer of streaming events on different platforms)*

> What was the big bug of it all? Economically it was not sustainable.
>
> *(S1., live streaming platform manager)*

Even the supposed successes mentioned earlier reveal several criticisms upon closer analysis. For instance, it is true that the Lacuna Coil concert gathered little more audience than it would have obtained in physical attendance, but tickets were sold at a much lower ticket price (€10), despite similar production costs, to which those for the streaming must be added. In the case of Heroes, since it was a charity event, the costs of streaming and the platform fee had been paid by the Ministry of Foreign Affairs, while the artists' fees had been greatly reduced; furthermore, a large part of the tickets had been purchased by corporate sponsors. In the case of Negramaro, the event was free and the costs were borne by the label, as it was a promotional event for the launch of the new album.

The problem of economic unsustainability is mainly explained by the relatively low number of participants, on which all the interviewees agree.

> Actually, live streamed concerts didn't work even during the pandemic [. . .] because the numbers were absolutely laughable, especially if free tickets, tickets linked to rewards or freebies of various types are removed from the evaluation of those (events) that went theoretically better, that is, if we consider only the effective ability to be able to charge for the ticket as the only way to enjoy the event.
>
> *(S2., collecting society for authors and publisher)*

Even extremely low prices have not produced the desired aggregation of a long tail of potentially interested people, and many interviewees pointed out that even free access was no guarantee of success:

> Thanks to some sponsors, we have even gone so far as to offer free streaming of famous artists in Italy, musicians who usually fill sports hall, however (even in these cases) we reached very few people, below a thousand, despite they didn't have to pay.
>
> *(C., live promoter and producer).*

At the beginning, the model envisaged small payments, it was based on the idea that the public was willing to pay a few euros to see this kind of content. So (the goal) was many very small payouts. Once we even started doing some free concerts, we realized that the price was not the biggest barrier: the main barrier was that the public seemed not interested.

(R., live promoter and former employed in a live streaming platform)

According to some interviewees, the only live streamed concerts that were truly a success were those in which a unique and exceptional event was designed ad hoc for streaming. However, these events would still have unsustainable costs for musicians without a consolidated global following, or rather that are mainly aimed at the Italian public.

There are three explanations most frequently provided by the professionals interviewed to explain the difficulties in attracting a larger audience, especially paying audience. The first obviously concerns the lower appeal of remote participation compared to being physically present in the space of the performance, as confirmed by the questionnaire; in particular, the most disappointing aspects of the live streaming events, for the respondents, were "not being in the same physical place as the musician" (66%) and "the lack of interaction with the performer" (45%), even more than the "poor physical involvement" (32%) and the fact of "not being around other participants" (30%).

The second explanation regards the adoption of the *pay-per-content* model, especially in the face of the large amount of similar content accessible for free. The third one refers to the increasing loss of the habit of enjoying live contents through digital platforms, compared to on-demand. Here again, the comparison with the home viewing of soccer is sometimes referred to, but in these cases to argue – in a rather approximate and superficial way – the scarce appeal of the live fruition of streamed concerts. The recurring consideration is that, in the case of concerts, live fruition cannot count on the incentive represented by the unpredictability of the result: while watching a match after it has taken place, presumably already knowing the result, significantly decreases the enjoyment, the same does not apply to concerts. This would imply that it makes no difference to follow a musical event as it unfolds or later, if available on-demand (something that would still be functional to its economic viability); however it does not explain why the impossibility of viewing it after it has taken place, as was the case for most of the events under discussion here, shouldn't constitute an incentive to watch it live. Furthermore, there would be many other differences to consider between these two types of events, probably more significant for understanding the different appeal of their fruition through live streaming, but such an analysis cannot be developed here due to space constraints.

If we look at both the considerations of the professionals and the data from the questionnaire, it would seem that the live streamed music events had intercepted a small portion of the audience who, not being able to attend the concerts, contented themselves with enjoying them in live streaming, also remaining substantially satisfied by virtue of aspects such as the comfort of home fruition and the reduced (or absent) expense, but it was still a portion too limited to sustain the offer.

Other criticisms reported by the interviewees concern some dysfunctional relationships between stakeholders. For example, according to platform managers and live promoters, a further obstacle to achieving economic sustainability was the cost and difficulty of obtaining the rights to make content available on-demand from the record labels. Besides, some interviewees, especially on the platform side, have pointed their finger at the lack of interest in live streaming on the part of most musicians, once the pandemic was over. Sometimes this statement has been related to a certain ostracism on the part of many live promoters, fearing that live streaming would drive part of the audience away from in-venue attendance, and to their influence on musicians' decisions.

Finally, it should not be underestimated the extent to which technical problems have produced negative publicity and discouraged the fruition of live streamed concerts: for example, on the occasion of the Ultimo concert on Live Now, which broke the ticket record, many people were unable to access for the very high volumes of users who connected at the same time, which caused a server crash. Live Now also suffered a lawsuit from the English PRS, the body collecting royalties for performing rights, for refusing to provide information about the revenue of some events.

4. Different Routes, Driven by Commercial Brands and Tech Giants

According to the professionals, there are two possible lines of development of streaming concerts. On the one hand, the extreme spectacularization, as in the case of music content created ad hoc to exploit (and promote) the most advanced devices (such as VR viewers), focusing on musicians with a huge global appeal. Obviously, these options require huge resources. On the other hand, content designed for a particular kind of emerging- or medium-level musician, who struggles to satisfactorily sell live concerts in physical venues but is very strong on social media, that is "whose main asset is not live performance" but follower engagement (O., live streaming platform manager). More specifically, this idea consists in designing

> formats with a strong storytelling component, in which musicians tell their story, for example about their album and the experiences from which it's born, while interacting with the public, something halfway between a concert and a meeting, supplemented by extra contents that precedes the event to build anticipation; for example, there could first be a content regarding the musicians in the studio, streaming live as they prepare the single and providing previews of the upcoming concert, or a content during the rehearsals for the concert; these contents could be left on-demand to create anticipation and engage people by accompanying them to the concert, which can be attended both in physical proximity and remotely".
>
> *(O., live streaming platform manager).*

In essence, this offering would no longer be about the individual concerts but would be conceived as a system of synergic contents articulating a transmedia narrative, of which the concert is the centerpiece or the end point. In this scenario, companies that can count on an integrated system of channels suitable for delivering different types of content would find themselves in an advantageous position, as in the case of Amazon, referred to for example by the manager quoted above: on the one hand it has Twitch, for contents focused on intimate storytelling and direct interaction between musicians and fans, on the other Prime Video, for contents in which the performance and the spectacular dimension are emphasized.

In the vision of the platform managers who put forward such ideas, these contents should be free. For this reason, a fundamental part of this scenario consists in the involvement of brands interested in the extremely engaged audiences of these musicians, through forms of collaboration that go far beyond the mere sponsorship of the event. According to the platform manager who deepened this idea the most, musicians would take on the role of influencers who produce 'live' content – including backstage and fan meeting – to which call-to-actions for purchasing goods can be anchored.

Such ideas are not something totally new, as they connect to another evolutionary line of the live streaming of musical events in the years before the pandemic. Actually, one of the most successful – and longest-lasting – live streamed concert series is represented by the branded content *American Express Unstaged*, launched as far back as 2010. Over the years, American Express Unstaged has streamed about thirty live concerts of global superstars, initially through partnerships with YouTube and VEVO, focusing on two further aspects to increase the appeal of the proposal: with regard to the artistic side, the direction of the concerts was entrusted to famous and loved filmmakers (e.g. David Lynch for Duran Duran, Werner Herzog for The Killers, Anton Corbijn for Coldplay, Terry Gilliam for Arcade Fire, Spike Lee for Pharrell Williams); with regard to new viewing experiences enabled by technology, spectators could choose their own view of the performance, send tweets to the bands, vote for the encore songs and interact with other viewers. Moreover, what is generally considered to be the most successful global music show streamed by the platform Live Now, Dua Lipa's *Studio 2054*, was in fact part of the American Express Unstaged series. Over the years the format has evolved in various ways: one of these was precisely the implementation of a more articulated storytelling, through the production of extra content involving the musicians and conceived as parts of a narrative structure, published before the concert in order to support its promotion. Another line of development concerned the experimentation of a spin-off dedicated to promising new acts (*Unstaged: Artists in Residence*): in this case, the format integrated a meeting between the musicians and social media promotion experts, designed to train the former "on how to best engage their fans while also gaining new ones", in addition to contents telling the stories of the musicians and the following concerts (McIntyre, 2014).

In the scenario imagined by some of the live streaming platform managers, streamed concerts should be free, or at least perceived as such, as in the case of '*Amazon Music Live*', which launched in October 2022 and brings live concerts of global stars to Prime Video subscribers. These events are usually streamed on both Prime Video and Amazon Music channel on Twitch, while some of them are also available on-demand for a limited time. Coincidentally, the first Amazon Music Live with an Italian artist, the singer Elodie, constituted, in May 2023, the first major live streamed musical event after a year in which no similar productions took place. The streamed event included a pre-show with famous guests, in which the singer's career was retraced. It is worth noting how Elodie largely corresponds to the profile outlined in the platform manager discourses quoted above: she has a large following on social media and her artistic trajectory was characterized by a growing presence on various media, culminating in several projects released in 2023. Elodie was initially known by the general public through her participation, in 2016, in one of the most popular Italian television talent shows (Amici), while in 2023 she was the protagonist of both a movie presented at the Venice International Exhibition, distributed also on Paramount+, and – above all – of a three-episode TV docuserie released, not so coincidentally, on Prime Video. The docuserie, entitled "Sento ancora la vertigine" (I still feel dizzy), takes viewers behind the scenes of her participation in the most important Italian televised musical event (the Sanremo Festival). In the words of a critic, the docuserie is built to arouse an impression of intimacy and authenticity, although "we are light years away from a real documentary, one that clearly transcends the promotional purpose or at least that of a storytelling organically created to conform to the whole communication of Elodie's image" (Giudici, 2023, my translation).

5. Conclusions

According to the industry stakeholders, the main limitations of the previous offer of live streamed concerts consisted – on the one hand – in an inadequate adaptation to the digital ecosystem (which penalizes *pay-per-view* content and favors on-demand fruition) and – on the other – in the mistaken belief that some innovative features would compensate for the loss of some of the appealing elements of face-to-face concerts, such as "being in the same physical space as the performer", something that was missed even to those who appreciated the customization of vision, the availability of meet and greet with the musicians or other interaction features.

The possible solutions, for those who try to imagine the development of live concerts in a digital environment, largely converge on the growing tendency to consider music as content for brand promotion (Meier, 2016) and some kind of musicians as influencers, especially those whose main value is considered to be the relationship with an engaged fandom, cultivated through interaction on social media.

Since the scenarios concerning the development of more sustainable models are linked to evolutionary coordinates preceding the pandemic, it almost seems that the latter has constituted not simply an acceleration in the development of the online concert offer as a diverging parenthesis to the prevailing ways in which the music business adapts to the digital environment. The pandemic has fueled unprecedented collaborations between new digital players, who have quickly entered the field to fill the sudden lack of concerts in physical proximity and – at least implicitly – to exploit the situation, and traditional players in the live sector, less accustomed to transposing their productions into a digital environment. The goal of implementing a digital alter-ego of concerts, which could initially function as their substitute and later develop in parallel, was largely pursued by retaining some of their production and economic logic and conceiving the opportunities for innovation mainly in relation to the modes of participation enabled by technology (again, however, often trying to simulate conventional modes of participation). Once these attempts have failed, the strategies to develop the offer of concerts in the digital environment seem to reconnect with the evolution of the music sector towards greater dependence on consumer brands and tech giants (Meier, 2016; Hesmondhalgh and Meier, 2018; Negus, 2019). In a digital economy that favors 'free' or advertising-subsized content, the economies of music – including streamed concerts – increasingly rely on its uses by both brands, aimed at gaining cultural relevance and promoting their products, and by tech companies, aimed at promoting the purchase of devices and services or collecting user data, ultimately at supporting other core businesses.

References

Cireddu, F. (2023). 'Reflecting on participation through livestreaming music events in times of pandemic' *Journal of World Popular Music*, 10(1), pp. 79–99.

Ferreira, F. L. *et al.* (2022). 'The rise and fall of live online music event consumption during the pandemic: an analysis from the perspective of practice theory', *Cadernos EBAPE.BR*, 20(3), pp. 401–416.

Giudici, E. (2023). *Elodie su Prime: perché la musica oggi si racconta con le serie*. Available at: www.roc kol.it/news-735500/elodie-serie-sento-ancora-la-vertigine-amazon-prime-video-recensione (Accessed: 3 August 2023)

Hesmondhalgh, D., and Meier, L. M. (2018). 'What the digitalisation of music tells us about capitalism, culture and the power of the information technology sector', *Information, Communication & Society*, 21(11), pp. 1555–1570.

Kjus, Y., Spilker, H. S., and Kiberg, H. (2022). 'Liveness online in deadly times: How artists explored the expressive potential of live streamed concerts at the face of COVID-19 in Norway' *First Monday*, 27(6). https://firstmonday.org/ojs/index.php/fm/article/view/12398/11103

McIntyre, H. (2014). American Express Highlights 3 About-To-Break Artists In New "Artists In Residence" Program. Available at: www.forbes.com/sites/hughmcintyre/2014/11/07/american-express-highlights-3-about-to-break-artists-in-new-artists-in-residence-program/ (Accessed: 3 August 2023).

Meier, L. M. (2016). *Popular Music as Promotion: Music and Branding in the Digital Age*. Cambridge: Polity Press.

Mouillot, F. (2022). 'The social and cultural dimension of 'platforming' live music: the case of the Hong Kong independent music scene during the Covid-19 pandemic', *Popular Communication*, 20(4), pp. 274–291.

Negus, K. (2019). 'From creator to data: the post-record music industry and the digital conglomerates', *Media, Culture & Society*, 41(3), pp. 367–384.

Onderdijk, K. E. *et al.* (2021). 'Livestream experiments: the role of COVID-19, agency, presence, and social context in facilitating social connectedness', *Frontiers in Psychology*, 12(647929), pp. 1–25.

Perez-Monteagudo, A., and Curras-Perez, R. (2022). 'Live and online music festivals in the COVID-19 era: analysis of motivational differences and value perceptions', *Review of Business Management*, 24(3), pp. 420–438.

Rendell, J. (2021). 'Staying in, rocking out: online live music portal shows during the coronavirus pandemic' *Convergence*, 27(4), pp. 1092–1111.

Vandenberg, F., Berghman, M., and Schaap, J. (2020). 'The 'lonely raver': music livestreams during COVID-19 as a hotline to collective consciousness?', *European Societies*, 23(sup1), pp. S141–S152.

4

YOU'RE NOT SUPPOSED TO LAUNDER MY MUSIC!

Music as Data in the Training of Generative AI Music Models

Rachael Drury

1. Introduction

The use of artificial intelligence (AI) in the generation of creative works was the hot topic of 2023. When ChatGPT (OpenAI, 2022) launched in November 2022, it brought AI to the public in a very usable way, and the power of AI became a widespread reality. In April 2023, Ghostwriter released the controversial song 'Heart on My Sleeve', featuring AI-generated vocals purporting to be high-value artists Drake and The Weeknd (Ingham, 2023). The track widely referred to as 'Fake Drake' made waves across the music industry, bringing the legal challenges of generative AI to the forefront of AI debates (Levine, 2023; Powell, 2023a; Tencer, 2023). The popularised term 'AI music' encompasses many uses of the technology across a variety of music making practices: creation, consumption, production, diffusion (Caramiaux and Donnarumma, 2021, p. 90) and performance. Generative AI, however, is more specific, referring to "deep-learning models that can generate high-quality text, images, and other content [including music] based on the data they were trained on" (IBM, 2023). The current generative AI music revolution brings powerful music creation tools and services to the hands of everyday people, regardless of their musical knowledge and expertise. Platforms such as MuseNet (OpenAI, 2019) and Boomy (2023) can generate ready-to-release multitrack songs and musical works at the click of a button, democratising the creative process to offer anyone entry to the industry. Crucially, generative AI music relies upon large quantities of high-quality human-created music, that may be copyright-protected, as data to train AI models to generate new music. The unauthorised use of copyright-protected music, and the fact that machines can generate music faster and at a far lower cost than humans, puts the livelihoods of music creators at risk by denying creators the right to licence their music to companies that may then replace them in the music market.

This paper explores the multifaceted implications of AI in the realm of music generation, focusing on the contentious play between AI-driven data mining and copyright law, and the impact of the practice of data mining on the music industry and music creators. Text and data mining (TDM) is the automated process of extracting and analysing large amounts of data to train AI models through information retrieval, pattern recognition, frequency distributions and predictive analysis. AI relies on TDM as a crucial component of its learning and decision-making processes, especially in tasks involving the analysis of large volumes of data. As AI applications become

DOI: 10.4324/9781003396550-4

increasingly data-driven, TDM remains a critical tool for harnessing the power of data for AI-driven solutions. In the context of this paper, TDM uses human-created music to train generative AI music models, effectively reducing copyright-protected music to mere data which may be considered copyright infringement. Following the Intellectual Property Office (IPO) (2020) consultation *Artificial Intelligence and Intellectual Property*, the United Kingdom (UK) government announced its intention to introduce a new copyright exception that would allow TDM for all purposes (Intellectual Property Office, 2022). Though lawful access to copyright-protected music would be a requirement, this new exception would leave music creators susceptible to having their music mined by AI music companies to train their models without the opportunity to refuse permission or receive payment for its use.

Jamie Njoku-Goodwin, former chief executive of UK Music, an umbrella organisation which represents the collective interests of the UK's music industry, commented that the government's proposal "would give the green light to music laundering" (UK Music, 2023a). Described by Njoku-Goodwin as "the bedrock of the UK music industry" (ibid.), licensing copyright-protected music is key to the economic viability of the music industry and to the income of music creators. The UK government's plan to amend copyright law would not only disrupt a copyright framework designed to protect human creativity but would threaten to undermine the business model of the music industry.

Following a brief overview of generative AI and UK copyright law, this paper will draw on extensive industry feedback to the IPO consultation, and insights from a survey of 141 UK music creators designed by the author, to consider three key issues: (1) the use of music as data; (2) TDM, intellectual property and copyright infringement; and (3) music industry and music creator responses to generative AI music and TDM. Furthermore, this paper considers prospective resolutions, including AI-driven metadata enhancements and audio watermarking, while stressing the importance of transparency and fair remuneration for creators. This study shines a light on the evolving landscape of AI music, advocating for communication and synergy between technology, creativity and copyright.

2.   Background and Context

2.1.   Generative AI and Music

The purpose of this paper is not to delve into the vast landscape of AI, but understanding how AI music is generated is fundamental to appreciating the significance of text and data mining (TDM) in the context of advancing generative AI music and its implications for copyright law.

Currently, generative AI systems are primarily based on machine learning (ML), a subfield of AI "that studies the phenomenon of learning" (Widmer, 2000, p. 70), and deep learning (DL), a "repertoire of machine learning (ML) techniques, based on artificial neural networks" (Briot, Hadjeres and Pachet, 2020, p. 1). Artificial neural networks (ANNs) are combinations of data structures and algorithms used to analyse data, inspired by the structure and function of the human brain. ANNs are DL processes that create "an adaptive system that computers use to learn from their mistakes and improve continuously" (AWS, 2023). They are typically comprised of multiple layers of nodes that extract features from data: the input layer of existing data, hidden layer(s) responsible for analysis, functions and transformations, and the output layer of new data. The depth of this architecture, with its multiple hidden layers, allows DL networks to automatically learn hierarchical representations of data, enabling them to tackle more complex functions and intricate patterns from data.

The key terms to note here are 'learning' and 'data'. Current generative AI systems based on ML and DL techniques are data-driven and utilise computational algorithms that have adaptive learning capabilities. ANNs depend on extensive datasets that contain complex data from multiple sources. These datasets, referred to as big data, require algorithmic operations to mitigate complexity and facilitate effective analysis. Various algorithms are employed, including statistical learning, pattern recognition, and probabilistic models, to generate new data that is similar to the dataset: this can now be achieved in a wide range of tasks, including image recognition, natural language processing, and music generation. The practice of extracting, analysing and learning from big data is known as text and data mining (TDM).

2.2. UK Copyright

Copyright in the UK is a legal framework and philosophy designed to protect the value of human creativity and to incentivise the result of human skill, labour and judgment while striking a delicate balance with the public benefits that can arise from the use and reuse of a copyright-protected work (Lilley, 2006). Without copyright protection, "there would be no incentive to produce music for economic ends" (Tschmuck, 2017, p. 57). At its core, copyright grants owners the exclusive right to copy, and the authority to require others to seek permission to do so, ensuring that they can benefit financially from their work. UK legislators were the first to introduce copyright protection for computer-generated works under the *Copyright, Designs and Patents Act 1988* (S. 9(3)): "the first copyright legislation anywhere in the world which attempts to deal specifically with the advent of artificial intelligence" (Young, 1987). This provides a unique vantage point to study the legal impact of AI on the creative industries.

To balance incentive with public good, over time the UK has amended copyright law to accommodate new forms of media and technological advancements, such as the advent of the Internet and digitisation. Amendments have been introduced to protect intellectual property, such as extending the duration of copyright, currently life of the author plus 70 years for musical and literary works (*Copyright, Designs and Patents Act 1988*, S. 12(2)). To balance protection with public good, several exceptions exist to allow limited use of copyright-protected materials without the need for permission or payment to the copyright owner (*Copyright, Designs and Patents Act 1988*, C. 3). Current copyright exceptions include the right to make copies for text and data analysis for non-commercial research (*Copyright, Designs and Patents Act 1988*, S. 29A) which balances the interests of rightsholders with public benefit.

At the time of writing, text and data mining (TDM) remains an exception in UK copyright law, so long as the user has lawful access to copyright-protected works and the practice of TDM is for non-commercial research (*Copyright, Designs and Patents Act 1988*, S. 29A). As such, should copyright-protected works be used as data for TDM by AI companies for commercial purposes, this would be considered copyright infringement if permission has not been granted by rightsholders, usually in the form of a purchased licence. The UK government began to explore the current legal implications of AI for copyright and creativity through the consultation *Artificial Intelligence and Intellectual Property* (Intellectual Property Office, 2020). As a result, the intention to amend copyright law to allow the use of copyright-protected materials without the need for permission or payment to the copyright owner for any purpose, including commercial use, was announced (Wagget, 2022). Coincidentally, this decision coincides with the government's aspiration for the UK to become a global AI superpower (Department for Business, Energy and Industrial Strategy, 2022). This could be construed as a power move to promote AI innovation within its borders by making UK copyright law more attractive to developers and investors.

Copyright has become essential to the economic framework of the music industry, where business models and the income of music creators are reliant upon the ownership, transfer and licensing of any copyright that subsists in a creative work. While the primary objective of copyright to balance interests suggests that we should consider flexibilities in the copyright framework, any disruption to this legal and economic framework could have vast consequences for the music industry and music creators.

3. Discussion

3.1. Music as Data

Should a creative work, such as music, be considered data? The UK's existing text and data mining (TDM) exception was introduced in 2014 for analysis of human-readable text (*The Copyright and Rights in Performances (Research, Education, Libraries and Archives) Regulations 2014*). This exception could be applied to a variety of tasks such as records management and document searches, or to address issues of unstructured data: TDM "enable[s] the use of works for research, such as using the text of magazine articles to identify a cure for malaria" (UK Music, 2022). As such, music industry responses to the IPO's consultation were strongly of the opinion that music cannot and should not be linguistically reduced to mere data (British Phonographic Industry, 2022; Ivors Academy, 2022; Music Publishers Association, 2022).

Music creators do not consider their musical works to be data. A self-selecting online survey, designed by the author and open throughout 2023, aimed to explore the awareness and opinions of UK-based music creators on potential copyright infringement and the UK government's emerging policy on AI, specifically the prospective amendments to copyright law. Purposive sampling (Trochim, 2006) was designed to identify participants for the study through various channels, including the author's own professional contacts, music industry organisations (e.g., the Incorporated Society of Musicians (ISM), Featured Artists Coalition (FAC)) and higher education music departments. The survey consisted of tick box responses and free text boxes to allow for both quantitative and qualitative data analysis. The survey results found that the majority of music creators surveyed in the UK would choose not to give AI companies permission to use their works as training data (see Figure 4.1) and would expect payment for any use of their works that was beyond their control through licence fees or royalties (see Figure 4.2).

There is however an alternative view that the process of digitisation effectively transforms musical works into dematerialised content and deintellectualised data: "in this context, digital works may be purely treated and dealt with as containers of data" (Borghi and Karapapa, 2011). The IPO has also since stated that data can be found in everything: "While we respect the creativity involved in copyright . . . there are datasets in all types of work" (Intellectual Property Office, 2022). If this is the case, then the progress of technologies and innovations previously adopted by the music industry have in fact enabled music to be considered data.

3.2. TDM and Intellectual Property

The potential unauthorised use of copyright-protected music for text and data mining (TDM) and AI training is one of the most controversial aspects of generative AI in the music industry. Currently, generative AI music models trained to generate new musical works are entirely dependent on the input of human-created works. While some AI music companies, such as AIVA (2024), appear to engage with ethical practices by employing human musicians to create datasets, by using music

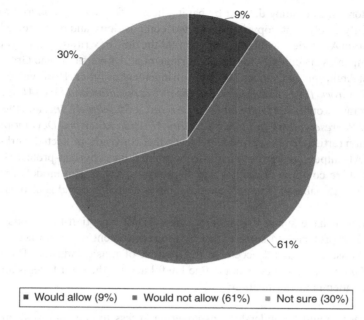

■ Would allow (9%)   ■ Would not allow (61%)   ■ Not sure (30%)

FIGURE 4.1 Creator perspective: music as data (permission to use music as data).

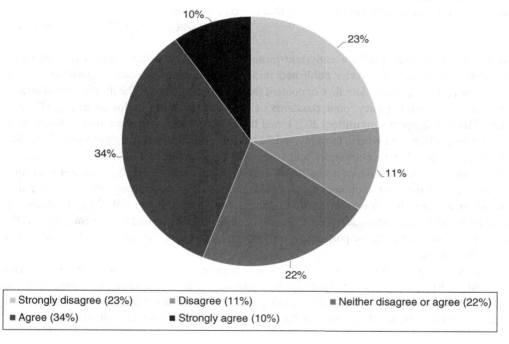

■ Strongly disagree (23%)   ■ Disagree (11%)   ■ Neither disagree or agree (22%)
■ Agree (34%)   ■ Strongly agree (10%)

FIGURE 4.2 Creator perspective: music as data (if remunerated).

in the public domain as training data, or by paying licence fees for any copyright-protected music used, it is highly likely that copyright-protected compositions and/or protected recordings are being used to train AI models without permission. At the time of writing, we are yet to see litigation against AI companies specifically for music infringement; Universal Music Group (UMG) filed a lawsuit against Anthropic in October 2023 for infringement of lyrics. However, emerging lawsuits such as *The New York Times Company v. Microsoft Corporation and OpenAI* in the realm of text generation (literary works) and *Getty Images (US), Inc. v. Stability AI, Inc.* and the class action suit *Individual and Representative Plaintiffs v. Stability AI, Midjourney and DeviantArt*, which explore image generation (artistic works) provide evidence that copyright-protected works are reproduced in generative AI outputs, therefore must also be trained on copyright-protected material. As AI music models share common principles and significant overlap with models for text and image generation, it is reasonable to assume that copyright-protected material is also used to train music models.

The government have stated that should a new TDM exception be introduced, the work of creators would be protected to some extent by the requirement of lawful access. However, it is unclear what constitutes lawful access in the context of music industry. The UK's *Copyright Designs and Patents Act 1988* does not define lawful access, though EU legislation (post-Brexit) has made some attempt at clarification:

> Lawful access should be understood as covering access to content based on an open access policy or through contractual agreements between rightsholders and research organisations or cultural heritage institutions, such as subscriptions, or through other lawful means . . . Lawful access should also cover access to content that is freely available online.
>
> *(Directive (EU) 2019/790,* Recital 14*)*

Researchers should only access a copyright-protected work if they have the legal right to do so through: a subscription; works published under open licences (Creative Commons, Open Government Licences, public domain); or content freely available online. The digitisation of music, the rise of digital online piracy, poor standards of remuneration from online streaming (Digital, Culture, Media and Sport Committee, 2021), and the cutthroat nature of the music industry have developed an environment where, to many, music has very little monetary value. Music is readily available and accessible to consumers and users for free through social media platforms and streaming services. In the wake of a new TDM exception for commercial use, a clearer definition of lawful access is required to prevent mass scraping of music as data through the Internet and to avoid potential disruption to current music distribution methods, should rightsholders choose to remove their music from these services. Crucially, clarification is needed to prevent a new TDM exception from undermining the purpose of copyright to protect the interests of rightsholders and motivate human creativity.

The proposed exception has been met with severe backlash across the creative industries, with only 13 out of 88 responses to the consultation in favour of such an exception. At the time of writing, the government has chosen to pause these plans, directing its attention towards the regulation of AI (Department for Science, Innovation & Technology and Office for Artificial Intelligence, 2023).

3.3. The Idea/Expression Dichotomy

Currently, AI companies are not acting lawfully if they are using copyright-protected music as training data; without permission this constitutes copyright infringement. However, the process of mining creative works for data is not considered to be copyright infringement if it is only factual information or ideas extracted from protected content (*Directive (EU) 2019/790*, Recital 9; Margoni and Kretschmer, 2022). This opinion stems from case law in the UK (*Donoghue v Allied Newspapers*, 1938) and the US (*Nichols v. Universal Pictures*, 1930) and from several directives in EU law. Only expressions of ideas can be copyright-protected, not the idea or factual information itself. This is known as the idea/expression dichotomy, a legal principle which attempts to set boundaries by using analysis to distinguish between protected expressions and unprotectable ideas or factual information. Determining infringement involves abstracting unprotected elements and comparing the remaining expressive elements for substantial similarity (Drury, 2021).

In the context of music creation, non-copyrightable ideas would include scales, arpeggios, chord progressions, and familiar combinations of rhythms, such as a bar of quavers. However, if it were the case that an AI was only extracting non-copyrightable elements of music, an AI could just be trained on music theory or music in the public domain. The fact that an AI model is being trained to mimic genres and styles of certain composers, songwriters and artists suggests that expressions of ideas are being copied and exploited when used as training data. Over The Bridge (2021) developed an album project called *Lost Tapes of the 27 Club*, which used AI to imagine what artists Amy Winehouse, Kurt Cobain, Jimi Hendrix and Jim Morrison might have created were they still alive. AI was used to generate new songs by analysing existing audio of the four artists, currently protected by copyright, to generate new music in their respective styles and genres. Though it is likely permission was granted by rightsholders for this project, it is an example of how the generation of popular music is dependent on training data from particular styles and genres or expressions of ideas.

3.4. The Berne Convention Three-Step Test

The UK is also bound by international copyright treaties such as the Berne Convention 1886, which attempts to harmonise the protection of works and the rights of their authors. The Berne Convention's three-step test, first enacted in the 1967 revision of the Berne Convention, attempts to standardise limitations and exceptions to exclusive rights by determining whether a use is 'fair'. Passing the three-step test allows users to make a partial or full copy of a copyright-protected work without permission. The Berne Convention states that:

> [I]t shall be a matter for legislation in the countries of the Union to permit the reproduction of such works in certain special cases, provided that such reproduction does not conflict with a normal exploitation of the work and does not unreasonably prejudice the legitimate interests of the author.
>
> *(Berne Convention for the Protection of Literary and Artistic Works, 1967,* Article 9*).*

The three-step test suggests that:

1. Limitations and exceptions cannot be overly broad ("special cases")
2. Limitations and exceptions cannot "rob rights holders of a real or potential source of income that is substantive ("conflicting with normal exploitation of the work")
3. Limitations and exceptions cannot "do disproportional harm to the rights holders" ("prejudice legitimate interests")

(Hugenholtz and Okediji, 2012).

As it stands, the UK's proposed text and data mining exception would fail to pass the three-step test. The exception would be overly broad to include all uses including commercial use; to sidestep licensing opportunities would deny rightsholders a potential source of income that is substantive. As the musical outputs of an AI model would be competing with the very works it is trained on, the exception could potentially do disproportional harm to rightsholders. The UK's proposed text and data mining exception would breach international copyright agreements.

3.5. The Implications of Amendments to Copyright Law

The music industry business model is almost entirely built upon the ownership, transfer and licensing of copyright. Music companies and music creators are heavily reliant on copyright to generate revenue and income. Any use of a copyright-protected work requires permission from the rightsholder in the form of a licence purchased through a collecting society, such as the Performing Rights Society (PRS). By denying the music industry the right to licence music to AI music companies, they are denying authors and rightsholders the right to remuneration for their creativity, as well as the right to prevent AI companies using their work without permission. Moreover, the government's proposal undermines the entire business model of the music industry.

The landscape of both generative AI music and copyright law has changed significantly in recent years. It is undeniable that the era of AI is at our doorstep as the next disruptive technology. As a threat to creativity, a trait so intrinsically human, AI is more concerning than digital piracy was at the turn of the century. AI has the potential to be the next mechanisation moment for the music industry: a time of technological change and structural economic reorganisation. This parallels the mechanisation of audio in the first half of the 20th century, marked by the innovation of the phonograph, radio, and talking pictures, which led to mass displacement of jobs among musicians (Kraft, 1996).

3.6. Music Industry Response

No stranger to disruptive technologies, the music industry is a champion of innovation, until it threatens to devalue the very model the business is built on. The music industry's response to AI has been somewhat chaotic, initially marked by calls for streaming services to prohibit AI companies from scraping content and to stop hosting AI-generated music (Gallagher, 2023; Nicolaou, 2023; Levy, 2023; Yurkevich, 2023). Universal Music Group also spearheaded the Human Artistry Campaign (2023), which delivers seven core principles outlining how artificial intelligence can be used responsibly to support human creativity with respect to the value of human artistry. One of these core principles states that the "use of copyright works and the use of voices and likeness of professional performers requires authorisation and free market licensing from all rightsholders".

While music creators do see some benefit to the development of AI, music creators are generally jaded with regards to the fight for fair remuneration at the hands of emerging technologies. They are struggling to see a way for human creators to be fairly remunerated for this type of use. In response to the UK government's plans to amend copyright law an anonymous survey participant commented:

> I would consider anyone profiting off someone else's creative work as exploitation and 'cause AIs can only learn from what's there already there's no way to have AI music without it sounding really outdated (from out of copyright music) or exploiting someone. I see it's inevitable though and to be fair music creators don't get a particularly good deal in any other aspect of their work so why should AI be any different?
>
> *(Survey participant: composer, producer, performer)*

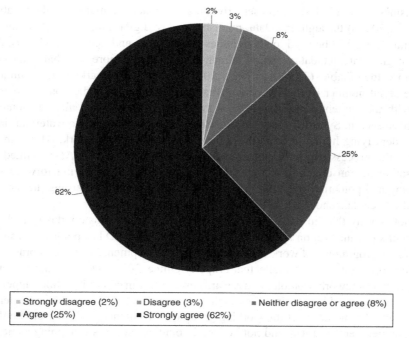

- Strongly disagree (2%)   - Disagree (3%)   - Neither disagree or agree (8%)
- Agree (25%)   - Strongly agree (62%)

FIGURE 4.3   Concern about government plans for copyright: a creator perspective.

Understandably, 87% of participants are concerned by the UK government's plan to allow AI companies to use human-created music without the permission of rightsholders (see Figure 4.3). Authors of creative works rightly place a high value on their work, but AI companies require hundreds of thousands of musical works to train their AI models. According to an AI company CEO, the value of Elvis would be a fraction of a cent to their company (Powell, 2023b, 33:41). Regardless, music creators feel that as AI music models are being trained on existing music, they are simply exploiting a human's creative work:

> AI music just creates a rip off of the thousands of songs used to 'train it'. It's directly stealing human created ideas. It's a machine and shouldn't get any praise for coming up with something 'good' because it didn't. It just combined thousands of existing ideas created by humans.
>
> *(Survey participant: composer, performer)*

Music creators are also concerned that the companies with a commercial interest in AI music have a stronger voice in discussion of law and policy than music creators, and that moves in favour of AI are inevitable:

> I imagine the commercial interest in groups advocating for UK government to allow AI companies to use human created music/songs have a disproportionately strong voice in these discussions in comparison to the vast body of human creators. This is part of the problem.
>
> *(Survey participant: student, composer, producer, performer, DJ)*

In general, music creators also feel that the genie is out of the bottle: it is going to be very difficult to regulate the use of human-created music to train AI models. It is agreed among music creators, music businesses, and industry representatives that AI-generated music should be labelled as such (UK Music, 2023b) through metadata, publisher codes, digital fingerprints or audio watermarks. In the future, AI could be used to solve issues of copyright and licensing. Already, AI can create multipoint granular metadata, making human-created music more searchable and AI-generated music more identifiable (ThinkAnalytics, 2023). Audio watermarking is a unique, invisible, inaudible serial number or marker pattern that can be added to audio content. It is designed to remain with the content regardless of how that content is used and altered. Some success has already been seen in SynthID (Google DeepMind, 2023). In time, audio watermarking could play a role in identifying human-created music used to generate an AI work. AI has the potential to streamline the process of identifying music used as training data in AI-generated works while compensating human creators for that use. For now, the key to moving forward with text and data mining and potential licensing opportunities is transparency regarding the use of copyright-protected works; data should be both identifiable and verifiable.

It is noteworthy that, in response to concerns raised by the music industry and as part of the government's pro-innovation approach to AI regulation, the IPO has published a summary of the government's programme of work. This includes the development of a code of practice specifically addressing copyright and AI (Intellectual Property Office, 2023). The code aims to ensure that the UK copyright framework promotes and rewards investment in creativity while supporting ambition for the UK to be a world leader in AI research and innovation. However, the agreement would only be entered into by members of the working group on a voluntary basis, it would only be adhered to by those members, and it would not be legally binding. So, it is a stepping stone rather than a solution. However, the code of practice aims to increase the accessibility of licences for TDM. This objective is geared towards helping AI companies in overcoming the barriers that they currently face, while ensuring robust protections for rightsholders. Once again, it addresses the need to strike a balance between the financial incentivisation of human creativity with public good. This could be the start of establishing a solid licensing framework for music as AI data which may negate the need for a text and data mining exception.

4. Conclusions

From a music industry and music creator perspective, this paper has discussed the significant implications of potential amendments to copyright law that would introduce a TDM exception for commercial purposes, allowing AI companies to mine music as data in the training of their AI music models. Generative AI music is wholly dependent on the quality and quantity of data that it is trained on, and AI models need tens, if not hundreds, of thousands of pieces of music in order to learn effectively. Any unauthorised use of copyright-protected music is an act of infringement if permission is not granted by the copyright owner, or if the use is not covered by a copyright exception. Music industry and music creators are concerned by the UK government's intentions to amend copyright in favour of AI innovation, over the interests of human creators who are still suffering the effects of the move to digitisation over two decades ago. With conversations ongoing regarding fair remuneration for music creators from streaming profits, it is vital that a good balance is found during the early stages of generative AI music development so that creators are incentivised to continue to create. Forming a dialogue between music creators, music businesses, AI companies, and policy and law makers is a good place to start.

As Jamie Njoku-Goodwin put it, a text and data mining copyright exception that would allow AI companies to mine music as data without the permission of rightsholders would give the green light to music laundering. Music laundering is happening right now in the name of progress and innovation. Music industry representatives and creators should work with AI developers to create tools that could be used to preserve creativity and copyright, and government should avoid blanket exceptions that undermine the work of human creativity and the entire business model of the music industry. AI could also be used to protect against music laundering and facilitate a fair licensing environment. Whether an infringer is human or artificial, infringement is infringement. Why should human creators be held accountable for copyright infringement if an AI cannot?

References

AIVA (2024) AIVA: your personal AI music generation assistant. Available at: www.aiva.ai (Accessed: 5 January 2024).

AWS (2023) What is a neural network? Available at: https://aws.amazon.com/what-is/neural-network/ (Accessed 5 January 2024).

Berne Convention for the protection of literary and artistic works [as amended on September 28, 1967]. Available at: https://wipolex.wipo.int/en/text/283698 (Accessed: 23 October 2022).

Boomy (2023) *Unleash your creativity: make music with Boomy AI*. Available at: https://boomy.com (Accessed: 5 January 2024).

Borghi, M. and Karapapa, S. (2011) 'Non-display uses of copyright works: Google Books and beyond', *Queen Mary Journal of Intellectual Property*, 1(1), pp. 21–55. Available at: https://papers.ssrn.com/sol3/papers.cfm?abstract_id=2358912 (Accessed: 22 October 2022).

Briot, J. P., Hadjeres, G. and Pachet, F.-D. (2020) *Deep learning techniques for music generation*. Switzerland: Springer.

British Phonographic Industry (2022) 'Submission of BPI (British Recorded Music Industry) Ltd to the IPO consultation on artificial intelligence and intellectual property' January 2022 [Consultation response]. Available at: www.gov.uk/government/consultations/artificial-intelligence-and-ip-copyright-and-patents (Accessed: 30 June 2023).

Caramiaux B. and Donnarumma, M. (2021) 'Artificial Intelligence in music and performance: a subjective art-research inquiry', in E. R. Miranda (ed.) *Handbook of Artificial Intelligence for music: foundations, advanced approaches, and developments for creativity*. Cham: Springer, pp. 75–96.

Copyright, Designs and Patents Act 1988. Available at: www.legislation.gov.uk/ukpga/1988/48/contents (Accessed: 3 April 2021).

Department for Business, Energy and Industrial Strategy (2022) *National AI Strategy*. Available at: https://gov.uk/government/publications/national-ai-strategy/national-ai-strategy-html-version (Accessed: 6 January 2023).

Department for Science, Innovation & Technology and Office for Artificial Intelligence (2023) A pro-innovation approach to AI regulation. Available at: www.gov.uk/government/publications/ai-regulation-a-pro-innovation-approach/white-paper#annexc (Accessed 29 August 2023).

Digital, Culture, Media and Sport Committee (2021) *Economics of Streaming*. London: Parliamentary Copyright House of Commons 2019. Available at: https://committees.parliament.uk/publications/6739/documents/72525/default/ (Accessed: December 5, 2021).

Directive (EU) 2019/790. Available at: https://eur-lex.europa.eu/legal-content/EN/TXT/PDF/?uri=CELEX:32019L0790&from=EN (Accessed: 8 November 2021).

Donoghue v. Allied Newspapers (1938) 3 Ch. D. 503.

Drury, R. (2021) A constellation of inconsistencies: questioning the blurred lines of music copyright infringement. Available at: www.iaspm.org.uk/iaspm/wp-content/uploads/2021/10/Rachael-Drury.pdf (Accessed: 3 September 2023).

Gallagher, W. (2023) 'The music industry wants Apple Music & Spotify to block AI training', *Apple Insider*. Available at: https://appleinsider.com/articles/23/04/12/the-music-industry-wants-apple-music-spotify-to-block-ai-music-training (Accessed: 15 April 2023).

Getty Images (US), Inc. v. Stability AI, Inc. (2023) United States District Court for the District of Delaware, case 1:23-cv-00135-UNA. Available at: https://docs.justia.com/cases/federal/district-courts/delaware/dedce/1:2023cv00135/81407/1 (Accessed: 5 January 2024).

Google DeepMind (2023) *SynthID*. Available at: https://deepmind.google/technologies/synthid/ (Accessed: 22 December 2023).

Hugenholtz, P. B. and Okediji, R. (2012) 'Conceiving an international instrument on limitations and exceptions to copyright', *Institute for Information Law Research – University of Amsterdam*. Available at: http://dx.doi.org/10.2139/ssrn.2017629 (Accessed: 22 October 2022).

Human Artistry Campaign (2023) *Core principles for artificial intelligence application: in support of human creativity & accomplishment*. Available at: www.humanartistrycampaign.com/ (Accessed: 3 September 2023).

IBM (2023) 'What is generative AI?', *IBM Blog*, 20 April. Available at: https://research.ibm.com/blog/what-is-generative-AI (Accessed: 3 January 2024).

Individual and Representative Plaintiffs v. Stability AI Ltd., Stability AI, Inc., Midjourney, Inc. and DeviantArt, Inc. (2023) United States District Court Northern District of California San Francisco Division, case 3:23-cv-002101. Available at: https://fingfx.thomsonreuters.com/gfx/legaldocs/myvmogjdxvr/IP%20AI%20COPYRIGHT%20complaint.pdf (Accessed: 5 January 2024).

Ingham, T. (2023) 'This AI Drake rip-off already has 250,000 plays on Spotify. How will the music industry respond?', *Music Business Worldwide*. Available at: www.musicbusinessworldwide.com/this-drake-ai-rip-off-alread-has-250000-plays-on-spotify-is-it-a-ticking-time-bomb-for-music-rightsholders/ (Accessed: 1 May 2023).

Intellectual Property Office (2020) *Artificial intelligence call for views: copyright and related rights*. Available at: www.gov.uk/government/consultations/artificial-intelligence-and-intellectual-property-call-for-views/artificial-intelligence-call-for-views-copyright-and-related-rights (Accessed: 24 March 2021).

Intellectual Property Office (2022) Artificial intelligence and intellectual property: copyright and patents: Government response to consultation. Available at: www.gov.uk/government/consultations/artificial-intelligence-and-ip-copyright-and-patents/outcome/artificial-intelligence-and-intellectual-property-copyright-and-patents-government-response-to-consultation (Accessed 22 June 2022).

Intellectual Property Office (2023) The government's code of practice on copyright and AI. Available at: www.gov.uk/guidance/the-governments-code-of-practice-on-copyright-and-ai (Accessed 30 June 2023).

Ivors Academy (2022) 'Annex – Response form' January 2022 [Consultation response]. Available at: www.gov.uk/government/consultations/artificial-intelligence-and-ip-copyright-and-patents (Accessed: 30 June 2023).

Kraft, J. P. (1996) *Stage to studio: musicians and the sound revolution, 1890–1950*. Rev. edn. Baltimore, MD: John Hopkins University Press.

Levine, R. (2023) 'The real problem with Fake Drake', *Billboard Pro*. Available at: www.billboard.com/pro/fake-drake-song-generative-ai-music-debate-problems/ (Accessed: 14 May 2023).

Levy, R. G. (2023) 'Why the power of AI may hold the key to distributing wealth throughout the music industry (Part 2)', *Music Business Worldwide*. Available at: www.musicbusinessworldwide.com/why-the-power-of-ai-may-hold-the-key-to-distributing-wealth-throughout-the-music-industry-part-2/ (Accessed: 18 June 2023).

Lilley, A. (2006) Inside the creative industries: copyright on the ground. Available at: www.ippr.org/files/images/media/files/publication/2011/05/inside_the_creative_industries_1527.pdf (Accessed: 3 September 2023).

Margoni, T. and Kretschmer, M. (2022) 'A deeper look into the EU text and data mining exceptions: harmonisation, data ownership, and the future of technology', *GRUR International: Journal of European and International IP Law*, 71(8), pp. 685–701. Available at: https://academic.oup.com/grurint/article/71/8/685/6650009 (Accessed: 15 October 2022).

Music Publishers Association (2022) 'Music Publishers Association AI open consultation questions' January 2022 [Consultation response]. Available at: www.gov.uk/government/consultations/artificial-intelligence-and-ip-copyright-and-patents (Accessed: 30 June 2023).

Nichols v. Universal Pictures (1930) 2nd Circuit, 45 F.2d 119. Harvard Law. Available at: https://cyber.harvard.edu/people/tfisher/IP/1930%20Nichols.pdf (Accessed: 28 January 2021).

Nicolaou, A. (2023) 'UMG tells Apple and Spotify to block AI lyric, melody scraping', *Arts Technica*. Available at: https://arstechnica.com/information-technology/2023/04/streaming-services-urged-to-clamp-down-on-ai-generated-music/?comments=1&comments-page=1 (Accessed: 15 April 2023).

OpenAI (2019) 'MuseNet', *OpenAI Research*, 25 April. Available at: https://openai.com/research/musenet (Accessed: 5 January 2024).

OpenAI (2022) 'Introducing ChatGPT', *OpenAI Blog*, 30 November. Available at: https://openai.com/blog/chatgpt (Accessed: 3 January 2024).

Over The Bridge (2021) Lost Tapes of the 27 Club. Available at: https://losttapesofthe27club.com (Accessed: 5 January 2024).

Powell, J. (2023a) 'The music industry freaked out over 'fake Drake'. But are these AI tracks just the latest take on remixing?', *Music Business Worldwide*. Available at: www.musicbusinessworldwide.com/the-music-industry-freaked-out-over-fake-drake-but-are-these-ai-tracks-just-the-latest-take-on-remixing/ (Accessed: 14 May 2023).

Powell, J. (2023b) Generative AI: exponentially impacting the arts. [Conference panel discussion]. The Namm Show, Anaheim Convention Center. 15 April. Available at: www.nammshow.org/event/the-2023-namm-show/planning/UGxhbm5pbmdfMTE4OTI3MA== (Accessed: 15 April 2023).

Tencer, D. (2023) 'AI 'fake Drake' track deleted on Spotify, YouTube, TikTok after Universal Music Group copyright claim', *Music Business Worldwide*. Available at: www.musicbusinessworldwide.com/ai-fake-drake-track-deleted-on-spotify-youtube-tiktok-millions-of-plays/ (Accessed: 1 May 2023).

The Copyright and Rights in Performances (Research, Education, Libraries and Archives) Regulations 2014 (SI 2014/1372). Available at: www.legislation.gov.uk/uksi/2014/1372/contents/made (Accessed: 5 January 2024).

The New York Times Company v. Microsoft Corporation and OpenAI (2023) United States District Court of New York, case 1:23-cv-11195. Available at: https://nytco-assets.nytimes.com/2023/12/NYT_Complaint_Dec2023.pdf (Accessed: 5 January 2024).

ThinkAnalytics (2023) *ThinkMetadata*. Available at: https://thinkanalytics.com/ai-driven-metadata/ (Accessed: 5 January 2024).

Trochim, W. (2006) *The research methods knowledge base*. 3rd edn. Ohio, OH: Atomic Dog Publishing.

Tschmuck, P. (2017) *The economics of music*. 2nd rev. edn. Newcastle upon Tyne: Agenda Publishing.

UK Music (2022) 'Artificial intelligence and IP: copyright and patents IPO consultation' January 2022 [Consultation response]. Available at: www.gov.uk/government/consultations/artificial-intelligence-and-ip-copyright-and-patents (Accessed: 30 June 2023).

UK Music (2023a) Music industry chiefs urge culture secretary to ditch "dangerous and damaging Artificial Intelligence plan. Available at: www.ukmusic.org/news/music-industry-chiefs-urge-culture-secretary-to-ditch-dangerous-and-damaging-plan-to-allow-artificial-intelligence-firms-to-data-mine-the-music-of-uk-creators/ (Accessed 3 January 2023).

UK Music (2023b) UK Music policy position paper on artificial intelligence. Available at: www.ukmusic.org/wp-content/uploads/2023/07/UK-Music-Policy-Position-Paper-on-Artificial-Intelligence.pdf (Accessed: 3 January 2024).

Wagget, D. (2022) 'UK government announces new text and data mining copyright exception in response to AI and IP consultation', *Dentons Blog*, 14 July. Available at: www.dentons.com/en/insights/articles/2022/july/14/uk-government-announces-new-text-and-data-mining-copyright-exception (Accessed: 3 January 2024).

Widmer, G. (2000) 'On the potential of machine learning for music research', in Miranda E. R. (ed.) *Readings in music and artificial intelligence*. New York: Routledge. pp. 69–84.

Young, D. (1987) 'Copyright, Designs and Patents Bill Hl', *Hansard: Lord's Chamber*, 12 November, 489, c. 1476. Available at: https://hansard.parliament.uk/Lords/1987-11-12/debates/9b959a7b-172a-4e28-8676-1a6747b0f370/CopyrightDesignsAndPatentsBillHl (Accessed: 3 January 2024).

Yurkevich, V. (2023) 'Universal Music Group calls AI music a 'fraud', wants it banned from streaming platforms. Experts say it's not that easy', *CNN Business*. Available at: https://edition.cnn.com/2023/04/18/tech/universal-music-group-artificial-intelligence/index.html (Accessed: 14 May 2023).

5

"UNCONSCIOUS REBELLIOUSNESS"

The Polytonal Roles of Young Musicians in an Innovative Music Program

Katy H. Weatherly

1. Introduction

As a performer of contemporary classical music, I have noticed a lack of understanding and appreciation for this genre among the general public and musicians across all levels of training. My personal journey with contemporary music began during my college years, where I initially had little experience or fondness for its sounds. However, through a "sink or swim" approach, I learned to embrace and appreciate contemporary repertoire, eventually becoming known for playing contemporary pieces in my recitals and juries at my college. Over time, contemporary music has become an integral part of my musical identity in adulthood, though its exact origin remains mysterious.

This sense of passion and intrigue surrounding contemporary music has fueled my interest in exploring the experiences of young musicians who share a similar fascination and appreciation for this genre among the general public and musicians across all levels of training. This sense of passion and intrigue surrounding contemporary music has fueled my interest in exploring the experiences of young musicians who share a similar fascination. I am particularly intrigued by the role contemporary music plays in shaping their musical identities, drawing inspiration from Simon Frith's idea that certain types of music become intertwined with our sense of self.

> The experience of pop music is an experience of identity: in responding to a song, we are drawn, haphazardly, into emotional alliances with the performers and with the performers' other fans. Because of its qualities of abstractness, music is, by nature, an individualizing form. we absorb songs into our own lives and rhythm into our own bodies; they have a looseness of reference that makes them immediately accessible.
>
> *(Frith, 1996, p. 121)*

I started pondering whether Frith's ideology could be extended to encompass contemporary classical music, which embodies a distinct "quality of abstractness." Contemporary classical music frequently takes on diverse forms, occasionally adopting the most abstract structures and tonalities. However, it is precisely due to this abstract nature that it frequently encounters misinterpretation and lack of recognition, both among the broader public and within the music community.

DOI: 10.4324/9781003396550-5

This curiosity has led me to focus on exploring the experiences of young string players involved in an innovative music program called *Face the Music*. This program caters to musicians aged 10 to 18 and is dedicated to preparing and performing music by living classical composers, emphasizing contemporary music. The purpose of this ethnographic research is to explore the experiences of young string players who are playing in a contemporary music ensemble collective through the lens of their identity. Through my investigation, I hope to shed light on the unique experiences and perspectives of these young musicians, exploring how they engage with contemporary music, what draws them to this genre, and how it shapes their musical identities.

1.1. *Contemporary Classical Music*

Defining contemporary classical music proves to be one of the most challenging aspects of this research. Despite my thorough examination of various dictionaries, a universally accepted definition for contemporary classical music remains elusive. Contemporary classical music, also known as avant-garde music, modern music, postmodernism music, or new music, is occasionally regarded as "high art." The exact beginning of the term in Western music history is unclear. It might have started in the search for different modes of expression by musicians and artists at the turn of the 20th century. The late romantic style promulgated by Gustav Mahler, Jean Sibelius, and Richard Strauss stretched the boundaries of post-romantic symphonic writing, while Claude Debussy and the art world pioneered Impressionism. Together, these artists created new uses of color, innovative chord combinations, and ambiguity in tonalities. Composers influenced each other yet moved in different directions, broadening the scope of unconventional tonality, introducing a "linguistic plurality" of styles, and pioneering innovative techniques and modes of expression (Morgan, 1984, p. 458). Béla Bartók used folk music to incorporate uncommon tonality in his works. Dimitri Shostakovich utilized a variety of different voices in his works, with elements of grotesque and ambivalent tonality.

In 1922, Arnold Schoenberg invented a 12-tone technique emphasizing atonality which was adapted by the composers Alban Berg, Anton Webern, and Hanns Eisler of the Second Viennese School. This technique was further explored by later composers like Pierre Boulez, Milton Babbitt, and Ernst Krenek. In the 1950s, R. Murray Schafer, John Cage, and Paulino Oliveros, in particular, pioneered post-war avant-garde music using electroacoustic music and implementing the non-standard use of musical instruments; others such as Milton Babbitt, Karlheinz Stockhausen, Luigi Nono, Krzysztof Penderecki, and Edgard Varèse expanded traditional concepts of sound and even noise. Pierre Schaeffer, who developed a unique avant-garde form known as *musique concrete*, featured sounds from musical instruments, voice, the natural environment, and synthesizers and computer programs. Edgard Varèse, Pierre Henry, Iannis Xenakis, Luciano Berio and more concentrated on electronic music and sound production.

Modernism in music is diverse and innovative, stretching well beyond the familiar and comfortable. Serialist, minimalist, experimental, electronic, postmodernist, new simplicity and new complexity music arose throughout the late 20th century and continued into the new millennium. To date, living composers have started to use a wider palette of musical composition techniques, mixing multicultural themes, multimedia, and a diversity of styles in their works. Some recent practitioners include Meredith Monk, Tan Dun, John Adams, Missy Mazzoli, and Ennio Morricone. As contemporary music repertoire has greatly expanded ever since the early 20th century, different projects in the United States have advocated a fuller range of musicianship by expanding music learning beyond just the classical and romantic periods.

According to Weber (2003), contemporary classical music carries dense and polemical meanings, creating a realm of high art where proponents and opponents find little common ground. Weber (2003) discussed that classical and contemporary music have historically been at odds. He described if one were to inquire with any avant-garde composer today about their experiences with symphony orchestras, they would likely encounter a barrage of resentment from the composers toward these institutions. On the other hand, if one were asking a subscriber to orchestra concerts about their opinions on the amount of new music performed, they would receive a contrasting reaction, which accuse composers of not caring about the public and writing solely for each other (p. 121).

To date, while an increasing number of concerts are incorporating contemporary classical music into their programs, there remains a noticeable scarcity of comprehensive research that truly delves into the domain of contemporary classical music education, particularly when it comes to non-professional musicians, specifically adolescents. Despite the growing presence of contemporary classical music within the music landscape, the educational aspects of this genre, especially as they pertain to younger individuals who engage in music for personal fulfillment rather than professional pursuits, remain significantly underexplored.

2. Related Literature

Contemporary classical music encompasses a vast and intricate realm of content, including technical and compositional terminology, subject to debate and possessing multidimensional facets. Moreover, contemporary classical music remains in a perpetual state of evolution and advancement. This literature review centers on two primary domains: (i) contemporary classical music education and (ii) the intersection of music and identity.

Contemporary classical music education has rarely been discussed, especially in the instrumental learning area for students. Many of the contemporary classical music research focuses on technicality, new innovations, repertoire analysis, but rarely on the education aspect for young students. Though recently, more research has been discussed for youths in the area of electronic music, hip-hop music, and popular music. Despite a lack of research on the educational aspects of contemporary music, especially for adolescents, it is worth mentioning the limited existing literature. Gordon (1950), Tuley et al. (1979), McGowan (1999), Auh and Walker (1999), and Allsup (2002, 2003, 2004) described multiple ways to introduce and teach contemporary music that allow students at various levels of familiarity and proficiency to understand context, overcome unfamiliarity, engage in divergent thinking, and increase creativity.

In the 1950s, Gordon (1950) delved into the intricate challenges that contemporary music posed for both music educators and students. He specifically examined two pivotal aspects of contemporary music—melody, harmony, and counterpoint marked by liberal use of dissonance and unconventional melodic intervals, as well as rhythm and meter characterized by the frequent employment of irregular meters and unusual rhythmic progressions (p. 38). What stood out was the realization that young musicians could not solely rely on reading the musical score to grasp the essence of the composition, as is often the case with classical music. To overcome this, Gordon proposed a multifaceted approach. He recommended an in-depth comprehension of the music, the ability to conduct irregular meters with precision, and the skill to place beats at the right moment within the music's structure (p. 39). Furthermore, Gordon underscored the significance of democratic teaching pedagogy in providing a more effective learning environment for students. This approach aimed to enhance the overall learning experience and understanding of contemporary music.

Tuley et al. (1979) presented comprehensive guidelines aimed at contemporary composers working on elementary-level music compositions. They advocated for a balanced approach that considers both mainstreaming and the ongoing emphasis on individualized learning within the open classroom setting. This, in turn, necessitates composers to craft a diverse repertoire that encompasses both non-traditional and traditional musical works (p. 47). The authors highlighted a notable issue—typical contemporary compositions for elementary students often lack supplementary instrumental components capable of accommodating the wide spectrum of student competencies. In response, they recommended that contemporary composers not only create fundamental, skill-graded supplementary parts for major instruments but also extend their efforts to include supplementary elements that cater to both ends of the competency spectrum. Additionally, a creative suggestion is put forth—to venture into composing music that seamlessly weaves together movement, dance, drama, voices, and instruments.

Turning to McGowan (1999), his insights revolved around the multifaceted nature of modern music's ambiguity and how students could better comprehend it. He highlighted the inherent associations and preconceptions tied to the term "Twentieth century music" (p. 21). McGowan shared his personal journey of growing fond of contemporary music and how it influenced his teaching at the college level. His perspective revealed that contemporary music wasn't a monolithic entity; rather, it encompassed a diverse array of styles spanning the past century. This diversity posed challenges in selecting appropriate repertoire and anticipating consistent audience reactions. To solve those challenges, McGowan offered pedagogical strategies to aid student understanding. He recommended exploring the evolution of musical styles from the 19th century, tracing how composers like Wagner, Debussy, and Stravinsky progressively intensified chromaticism, leading to Schoenberg's revolutionary 12-tone system. McGowan proposed introducing students to accessible "bridge" composers like Bartók, who could serve as stepping stones to comprehending more intricate works by other new music composers. Importantly, McGowan emphasized that the objective of teaching contemporary music wasn't merely fostering appreciation; it also aimed to nurture sensitivity to crucial musical elements such as rhythm, meter, timbre, and non-pitched elements.

Auh and Walker (1999) explored graphic notation's impact on young students' musical creativity: 38 students in their 7th grade in Seoul, compared graphic notation to traditional staff notation. Students were randomly assigned to create short musical pieces using either notation. Performances were recorded, and students explained their compositional strategies via a questionnaire. Evaluation based on Musical Originality, Musical Syntax, and Artistic Sensitivity revealed that non-traditional compositions showed higher musical creativity due to diverse strategies. Allsup (2002, 2003) engaged students in small groups, offering activities that covered both classical and popular composition approaches. Through studying their compositional process in connection with community dynamics, he concluded that the chosen music genre directly influences the students' community-building experience. Notably, a group that initially pursued classical composition became isolated; communication was poor, and compositional progress was limited. Consequently, they shifted to jazz and rock, fostering collaborative composition within a democratic and mutually enriching context. Consequently, Allsup advocates for an innovative approach—an instrumental ensemble that merges traditional pedagogies with popular music practices.

Given music's significant impact on adolescent development, it becomes crucial to delve into their identity formation, musical preferences, and identities. Music finds its place in various personal and social contexts. A comprehensive study conducted by North, Hargreaves, and Hargreaves (2004) involved 346 participants who participated in a 14-day survey about their music engagement. The results unveiled that while many music-listening instances occurred with others, individual enjoyment for music heard alone surpassed that experienced in a social setting. Classical and jazz music were less frequently encountered. For the majority, music was a leisure pursuit, yet seldom the primary focus. This study concluded that individuals consciously employ music in diverse interpersonal and social contexts to evoke distinct psychological states. The context shapes musical experiences and dictates their significance (p. 75).

From a sociological perspective, identity is more than just self-identification (O'Brien, 1995); it changes and transforms, and it is not a singular unit (Burke, 2006). Hudak (1999) posits that music serves as a dynamic, consciousness-altering medium, capable of shaking the foundations of daily life and paving the way for a shared realm where differences can intermingle and interact (p. 447). Hudak envisions the musical process as fostering a sense of collective identity—a musical "We" (p. 447). This 'We' establishes a community of longing, akin to bell hooks's concept wherein shared yearning cultivates a space where differences converge (p. 13). Hudak extends this idea by using Schutz's (1951) concept of inner time alignment to illustrate the relationship between composers and performers. Performers become the stewards of harmonious resonance with the composer, forging a connection of shared wavelength (p. 453). Hudak asserts that music-making forms a distinctive temporal structure, intrinsically linked to the essence of human existence. Crafting a sonic identity encompasses this intimate dimension; without it, human presence in social connections would be incomplete (p. 468).

According to Frith (1989), identity is depicted as fluid and transformative, involving continuous evolution rather than a static state. Similarly, our engagement with music, encompassing creation and listening, mirrors this ongoing self-development. Music, like identity, serves as both a performance and narrative, connecting the personal with the societal and the mental with the physical. Just as music possesses ethical and aesthetic dimensions, so does identity. Drawing inspiration from Frith's concept of identity, I embarked on an exploration of the young musicians associated with *Face the Music*.

3. Methods

This study adopts an ethnography-based single-case design, defined by Yin (2009) as an empirical investigation into a contemporary phenomenon within its real-life context, particularly when the boundaries between the phenomenon and context are unclear (p. 18). Single-case studies, as noted by Siggelkow (2007), offer a detailed portrayal of phenomena. Stake and Trumbull (1982) further emphasized that the purpose of a single-case study is not to establish generalizations but to capture the distinctiveness of the case. I employed diverse data collection methods, including semi-structured interviews, focus group interviews, observations, and field notes. I utilized multiple techniques to analyze nonverbal cues, emotions, and interactions between young musicians and teachers. Additionally, I observed their responses to specific musical pieces and rehearsal strategies. For instance, I closely observed how young musicians collaborated, which unveiled an emergent theme of community within *Face the Music*. Field notes were taken to capture both descriptive and reflective information, providing valuable insights and themes to shape the study. I also engaged with quartet members to delve into their group dynamics and distinct personalities.

Participants encompassed both young string players, termed "young musicians" in this paper, all of whom had a connection with *Face the Music*. The selection criteria were straightforward: participants needed to be involved with *Face the Music*. Young musicians, ranging from 10 to 18 years old, played various instruments at different proficiency levels and brought diverse educational and musical experiences. Some attended public schools while others were enrolled in private institutions. Throughout my data collection, I interviewed five focus groups of students, each containing three to four students—a total of 18 students for the focus groups as well as the founder. In this paper, pseudonyms were used for young musicians.

3.1. The Research Site: A Youth Innovative Music Program

In recent times, the United States has witnessed the emergence of innovative music programs tailored for youth. Among these, *Face the Music* stands as a unique entity. Unlike conventional classical music initiatives, *Face the Music* holds the distinction of being North America's sole program exclusively dedicated to the exploration and rendition of compositions by contemporary living composers, a category that includes its own members. Every Sunday, students ranging from 10 to 18 years old convene to engage in activities such as improvisation, jazz, composition, music technology, and the incorporation of cutting-edge technologies like virtual reality. According to *New York Magazine*, *Face the Music* cultivates a distinctive subculture where the realms of Radiohead and Portishead intersect with those of Bartók and Dvorák, effectively revitalizing orchestral music as a means of personal expression.

Originating in 2005 as an experimental afterschool club for a New York City public school, *Face the Music* has evolved into a well-established program. Its primary objective revolves around the practice and performance of compositions by living composers. The concept of directly collaborating with composers spurred this focus on contemporary repertoire. Recognizing their exceptional dedication and passion for championing the music of living composers, the *American Society of Composers, Authors and Publishers* (ASCAP) awarded the founder and *Face the Music* the prestigious *Aaron Copland Award* in 2011. The program encompasses diverse offerings, including string quartet programs, composition labs, a modern big band, and an orchestra. Special projects abound, wherein students assemble into distinct groups to rehearse and present performances within their concert series. Throughout the school year, this concert series comprises an impressive array of up to 40 performances, ranging from formal concert halls to local venues, galleries, school auditoriums, and even experiential performance spaces. As of now, *Face the Music* boasts approximately 200 members participating in multiple ensembles, embodying a thriving and multifaceted musical community.

4. Findings and Discussions

In my pursuit of unraveling young musicians' identities, I dedicated considerable time to engage in observations and conversations with them. Through these interactions, I inquired about a variety of topics, such as their musical preferences, their parents' viewpoints, and their perceptions of themselves within *Face the Music*. From this exploration, two primary findings emerged that I wish to highlight in relation to the musical identities of these young musicians: (i) "Unconscious Rebelliousness" and (ii) "Polytonal Roles."

A common theme that emerged between two of the quartet groups was the seeming lack of parental support for their children's contemporary music education. Despite this outright lack of support, parents continued to pay the dues for their children's enrollment, some for many years. The lack of enthusiasm shown by the parents contrasted starkly with that of their children. This made me wonder whether or not these young musicians could be considered rebellious and if they were aware of this possible rebelliousness, or perhaps an inner sense of autonomy and exploration? When I asked about their parents' thoughts on the music they performed at *Face the Music*, they answered:

Emily: My mom is very into classical music, and sometimes she really enjoys what we play in *Face the Music* and sometimes she really didn't like it.

Mike: My dad can't stand *Face the Music*, he likes all classical music . . .

Caroline: Well, my mom was against the fact that I am going to *Face the Music* . . . "I don't want you to continue going in." But for that, I just kept going in being in *Face the Music* for the entire high school years.

Numerous researchers (Creech & Hallam, 2003; Davidson et al., 1996; Zdzinski, 1996) have underscored the significance of parental encouragement in students' musical pursuits and instrumental studies. While parents exhibited support for their continued instrument playing, their endorsement for participation in *Face the Music* was limited, largely due to the unfamiliar repertoire and tonalities the program presented. Nonetheless, several young musicians attended *Face the Music* despite lacking parental backing. This prompted me to delve into the realm of potential rebelliousness among them. During the subsequent phase of my field observations, I posed a question to the young musicians: Did they view themselves as "rebellious," potentially deviating from their parents' wishes and engaging in something markedly different?

Caroline: I am just playing my violin; how does this make me rebellious?

Interestingly, none of the young musicians identified themselves as rebellious. Given their unwavering commitment to *Face the Music*—attending rehearsals punctually, never missing a session, and practicing diligently at home—how could they be considered rebellious? On one hand, I firmly believe that these young string players within *Face the Music* are inherently self-motivated. Some continue their involvement in the program even in the absence of parental or peer approval. On the other hand, their comments suggest a potential "unconscious rebelliousness." Ter Bogt et al. (2011) propose that adolescents, through their affinity for bold and unconventional music, establish a sense of independence from parental authority (p. 301). I introduce the term "unconscious" due to their possible lack of awareness regarding their implicit stance. Research also underscores that healthy parent-adolescent relationships foster autonomy (Grotevant & Cooper, 1985). Given the mutual respect shared between parents and the young string players in *Face the Music*, the latter may perceive themselves more as "independent" than "rebellious." Although they choose to engage in music that diverges from their parents' preferences, they diligently work toward becoming more accomplished musicians—a goal cherished by every parent at *Face the Music*.

Hesmondhalgh (2008) stated the major musical genres of the previous century —jazz, rock, soul, and hip-hop—have all been closely intertwined with idealized concepts of personal independence. Rock music, in particular, traversed historical shifts as described by Boltanski and Chiapello, generating a culture centered around *rebellious creativity*. However, in hindsight, it swiftly became associated with commercial values. The mainstream rock music of the 1980s and 1990s, often marked by its simplistic celebrations of freedom and unbridled individualism, aligns remarkably well with Boltanski and Chiapello's interconnected worldview (p. 335). However, classical music and traditional Western musical instruments often carry a contrasting ideology, labeling the young musicians in such bands as "band geeks." Now, *Face the Music* is this fusion place that combined the "band geeks" with the new music. Here, I argue that young musicians without this *unconscious rebelliousness* would not join *Face the Music*. In fact, the founder of *Face the Music*, Jenny Undercofler, expressed to me in the interview that she believed being in *Face the Music* was not exactly cool, but *weird*.

Jenny: Well, I saw friendships forming within the group. I don't think they were always regarded as cool by their peers; in fact, I knew that they weren't. Being in *Face the Music* is *weird*.... Well, listen, I am very much at the outset that these were the students that already decided they were not going to necessarily succeed in the traditional way. And so it was fun to see them succeed in a nontraditional way. And discovered that they could do things that they didn't feel they could. they could do things that they didn't feel they could. And to me, it was a huge piece of what felt like success to me.

Hence, I employed the term "unconscious" to symbolize this duality within their identities. These young musicians did not fit the "cool" archetype; they were perhaps labeled as "geeks" yet did not align enough with the conventional image of success in classical music. Thus, they found themselves in *Face the Music*, even in the face of limited support from their parents. However, this "unconscious rebelliousness" is precisely what set them apart, channeling innovation from the upcoming generation. If they had followed a mainstream "rebellious" path, they might have leaned towards rock, jazz, indie, or electronic music. Yet, their "unconscious rebelliousness" distinguished them as a distinctive presence within the intricate realm of music.

Another finding of the research would be the "polytonal roles" of these young musicians. While desperately searching these young musicians' identities, I found that many of them were unable to define their musical roles.

Christine: I don't think we are just playing the music. [But] I don't really know what "music is representing us" means.

I contend that the young musicians engaged with *Face the Music* exhibit a higher degree of open-mindedness, rendering them more inclined to embrace risk-taking and novel concepts. I perceive these individuals as characterized by a form of identity-driven thinking that promotes adaptability and receptivity to change. This resonates with Bolton and Reed's (2004) notion of "identity-driven thinking," which can lead to biased judgments resistant to change—a procedural bias or "sticky prior" rooted in initial identity-based judgments (p. 398). Many among the young musicians in *Face the Music* might not define themselves within a single identity since they encompass more than just their musical roles. Their participation in *Face the Music* represents merely a brief engagement within their busy schedules. It was not unexpected that they might hesitate to label themselves while they are still exploring various dimensions of their identities. When asked about their potential pursuit of a music major, many expressed interests, yet also mentioned a curiosity for alternative paths. While they remain receptive to fresh tonalities, ideas, and experiences, numerous young musicians have revealed interest in fields beyond music. Their identities within *Face the Music* comprise just a segment of their multifaceted selves, one that evolves over time. This "polytonal roles" aligns with the notion of "Unconscious Rebelliousness," reflecting their inclination to not conform solely to preconceived expectations.

The hierarchal self-concept implies that the judgments, cognition, and behaviors of one person tend to be consistent with the identities that are more important to that person. One's self-concept is a collection of beliefs, which comprise self-schemas that interlock with self-esteem, self-knowledge, and the social self. These self-identifications also involve past, present, and future selves (Ayduk et al., 2009; Myers, 2009). Not only does self-identification involve a multitude of timeframes, but it is also contradictory and fluid. Recent scholarship has explored the complexities of the relationship among the multiple dimensions of self, that is, the interpersonal, the intrapersonal, and the cognitive (Jones & McEwen, 2000). As young individuals explore their identities, they frequently lack the vocabulary or awareness to neatly fit themselves into the categories that adults often seek. I term these identities as "Polytonal Roles." Aligning their identities with the essence of contemporary classical music, which is often intricate, dynamic, complex, non-singular, and polytonal in nature.

5. Conclusions

Both the "unconscious rebelliousness" and "polytonal roles" exhibited by young musicians has the potential to be the catalyst for fostering innovation in the next generation. Innovation thrives when individuals challenge norms and venture into uncharted territories. Unlike conventional rebellion, this "unconscious rebelliousness" permits genuine exploration of fluid and ever-evolving identities. In my perspective, *Face the Music* serves as a realm of ambiguity—a sanctuary capable of embracing the intricate, polytonal nature of self-conceptions, akin to the complexity found in contemporary music.

If we perceive one's role as an unchanging, singular construct, it confines potential and hinders possibilities, particularly for early adolescents and youths. What set this environment apart was its acceptance of "unsureness." In modern education, the embracing of fluidity, complexity, and ambiguity often falls short. The focus tends to be on the "black and white," sidelining the valuable "in-between." During youth development, unconventional rebellion and individuality might be met with skepticism, yet they hold the potential to be essential qualities for driving innovation. I firmly believe that equipping young musicians with insight and admiration for contemporary music will not only enhance their musical journeys but also play a pivotal role in fostering a wider embrace and acknowledgment of this genre within the classical music realm. To foster innovation, educational programs should explore the integration of traditional instrumental education with contemporary performance practices. By embracing both sides, a more comprehensive and adaptable approach to musical education can be established.

References

Allsup, R. E. (2002). *Crossing Over: Mutual Learning and Democratic Action in Instrumental Music Education*. Teachers College, Columbia University.

Allsup, R. E. (2003). Mutual learning and democratic action in instrumental music education. *Journal of Research in Music Education*, *51*(1), 24–37.

Allsup, R. E. (2004). Imagining possibilities in a global world: Music, learning and rapid change. Music Education Research, 6(2), 179–190.

Auh, M. S., & Walker, R. (1999). Compositional strategies and musical creativity when composing with staff notations versus graphic notations among Korean students. *Bulletin of the Council for Research in Music Education*, 2–9.

Ayduk, O., Gyurak, A., & Luerssen, A. (2009). Rejection sensitivity moderates the impact of rejection on self-concept clarity. *Personality & Social Psychology Bulletin*, 35, 1467–1478.

Bolton, L. E., & Reed, A. (2004). Sticky priors: The perseverance of identity effects on judgment. *Journal of Marketing Research*, *41*(4), 397–410.

Burke, P. J. (2006). Identity change. *Social Psychology Quarterly*, *69*(1), 81–96.

Creech, A., & Hallam, S. (2003). Parent–teacher–pupil interactions in instrumental music tuition: a literature review. *British Journal of Music Education*, *20*(1), 29–44.

Davidson, J. W., Howe, M. J., Moore, D. G., & Sloboda, J. A. (1996). The role of parental influences in the development of musical performance. *British Journal of Developmental Psychology*, *14*(4), 399–412.

Frith, S. (1989). Why do songs have words?. *Contemporary Music Review*, *5*(1), 77–96.

Frith, S. (1996). Music and identity. *Questions of Cultural Identity*, *1*(1), 108–128.

Gordon, P. (1950). Rehearsing contemporary music. *Music Educators Journal*, *37*(1), 38–40. Retrieved from www.jstor.org/stable/3387286

Grotevant, H. D., & Cooper, C. R. (1985). Patterns of interaction in family relationships and the development of identity exploration in adolescence. *Child Development*, *56*, 415–428.

Hesmondhalgh, D. (2008). Cultural and creative industries. In T Bennett & J Frow (Eds.), *The SAGE Handbook of Cultural Analysis*, 552–569. SAGE Publications Ltd, https://doi.org/10.4135/9781848608443

Hudak, G. M. (1999). The "sound" identity: music-making & schooling. *Counterpoints*, *96*, 447–474.

Jones, S. R., & McEwen, M. K. (2000). A conceptual model of multiple dimensions of identity. *Journal of College Student Development*, *41*(4), 405–414.

McGowan, D. (1999). Teaching modern music. *American Music Teacher*, *49*(1), 21–24. Retrieved from www.jstor.org/stable/43545229

Morgan, R. P. (1984). Secret languages: The roots of musical modernism. *Critical Inquiry, 10*(3), 442–461.

Myers, D. G. (2009). *Social Psychology* (10th ed.). McGraw-Hill Higher Education. ISBN 978-0073370668.

North, A. C., Hargreaves, D. J., & Hargreaves, J. J. (2004). Uses of music in everyday life. *Music Perception, 22*(1), 41–77.

O'Brien, L. F. (1995). Evans on self-identification. *Nous, 29*(2), 232–247.

Schutz, A. (1951, March). Making music together-A study in social relationships. *Social Research, 18*(1), 76–97.

Siggelkow, N. (2007). Persuasion with case studies. *Academy of Management Journal, 50*(1), 20–24.

Stake, R. E., & Trumbull, D. J. (1982). Naturalistic Generalizations," *Review Journal of Philosophy and Social Science, 7*(1 & 2), 1–12.

Ter Bogt, T., Delsing, M., Van Zalk, M., Christenson, P., & Meeus, W. (2011). Intergenerational continuity of taste: Parental and adolescent music preferences. *Social Forces, 90*(1), 297–319. Retrieved from www.jstor.org/stable/41682642

Tuley, R. J., Rentschler, D. M., Pardue, T., Banks, S., Calvin, E. B., Jothen, M., & Garner, F. G. (1979). Point of view: Guidelines for contemporary composers of elementary-level music. *Music Educators Journal, 65*(7), 47–49.

Weber, W. (2003). Consequences of canon: The institutionalization of enmity between contemporary and classical music. *Common Knowledge, 9*(1), 78–99.

Yin, R. K. (2009). *Case Study Research: Design and Methods* (Vol. 5). Sage.

Zdzinski, S. F. (1996). Parental involvement, selected student attributes, and learning outcomes in instrumental music. *Journal of Research in Music Education, 44*(1), 34–48.

6

INNOVATION IN MUSIC

Reimagining Approaches to Gender Equality in the British Live Music Industry

Corinna Woolmer

1. Introduction

Gender inequality in the British music industries is an historic, ongoing issue which has been widely documented (Gill, 2002; Henry, 2009; Raine and Strong, 2018). Long before the Covid-19 pandemic, problems existed in the British music industries that cannot be ignored, and which, to some extent, the pandemic and the immediate effects generated by it have served to highlight. For example, the music industries have certainly not been immune to the highly gendered division of labour practices identified within other creative practice arenas (Conor et al., 2015; Reimer, 2016); a pronounced inequality of opportunity and progression surrounding gender exists in live music both in relation to artists on stage (Houston, 2021) and to those working backstage (Malt, 2021). Meanwhile, the anger at the toxic sexist culture long prevalent within the worldwide music industries (Scenestr, 2021), including the British live performance sector, is now being voiced like never before. Gender inequality manifests in a multitude of guises and spaces within the music industries, and as such, there is varied research and literature across a range of areas of gender, as well as other inequalities, faced within the industry. Much of the existing research and literature surrounding gender addresses the issue of misogyny and sexism in music and often focuses on musicians, music creators or music production, such as: sexist or misogynistic lyrics (Adams and Fuller, 2006; Cobb and Boettcher, 2007), particularly in rap music (Weitzer and Kubrin, 2009); how music and popular culture normalise violence against women (Hill et al., 2020); gender inequality of performers, such as how female artists (Wang and Horvát, 2019), music creators (Berkers et al., 2019) and screen composers (Cannizzo and Strong, 2020) are significantly under-represented in the music industries; how the construct of the music industries has allowed violence against women to be ignored (Strong and Rush, 2018); and gendered politics and misogynistic attitudes towards female stars (Williamson, 2010). There is also research relating to the normalisation of the mistreatment of women in and by the music industries towards audience members, industry workers and artists (Hill et al., 2020), as well as the harassment experienced by those working in the industry where Brown (2021) found through their investigation of the live music industry initiative *Safe Spaces In Music* that "over 60% of (music sector) workers have experienced sexual harassment". It is also widely reported that women are grossly under-represented in the music industries both as artists: in the recording industry (Bayton, 2003; Samuelson, 2019); and on stage

DOI: 10.4324/9781003396550-6

(Bain, 2019), where only 21% of acts at UK music festivals in 2021 were female (Gallop, 2021), as well as those working off-stage (Niethammer, 2019); and reports of female music artists earning less than male stars are plentiful (Lieb, 2018) which also extends to those working behind the scenes (Stassen, 2019) and the music industries as a whole (Bain, 2019). Additionally, there are other pressing ethical and social issues present within the industry that many feel the sector has been slow to respond to. These include the pay and conditions of those working within the music industries (Bradshaw, 2019), and other matters of welfare for those working in live music, including the relative neglect shown towards the high levels of mental health illness experienced by staff (Awbi, 2017). This article focuses on just one social and ethical issue in the area of inclusivity and equality in the British live music industry: gender. And specifically, pursuing workable solutions for the improvement of gender equality and inclusiveness in the British live music industry, by identifying and breaking down barriers faced by women.

2. Background to Gender Inequality in the British Live Music Industry

The British and indeed, global live music industries, have a well-documented and long-standing gender inequality problem. While such inequalities can manifest in a multitude of ways, the overarching problem of gender inequality was created, and then sustained, by the building of barriers which hinder and/or prohibit women's access into the industry (Coleman, 2020; Wattis et al., 2006), and limit the advancement of those women who do succeed in entering the industry (Raine and Strong, 2021; Vandello et al., 2013). Barriers are defined by Boot et al. (2018) as a factor, state or condition that limits, prohibits or prevents a person from either access or advancement (Oluyede et al., 2022). Gender inequality in the British music industries has been widely researched and documented, including recently, by UK Music (2022) and Incorporated Society of Musicians – ISM (2018, 2022). But relatively little research has detailed examples of successful gender equality measures in either the music industries or in parallel sectors. In particular, little to no attempt has been made in existing studies to investigate which facilitators, as defined by Boot et al. (2018) as factors that potentially facilitate, encourage or enable a person's access or advancement, might prove most impactful for improving women's access to the British live music industry, how they might be implemented into industry organisations and then for creating equality of advancement within it. In other words, existing work on this topic has often clearly mapped the problem of gender inequality in the British music industries but has offered little by way of suggestions for how the situation might be improved. But at a time when the music industries are aiming to build back better, in the wake of the Covid-19 pandemic as outlined by UK Music (2021), there may be no better opportunity to explore facilitators for improved gender equality in organisations (Tabbush and Friedman, 2020), especially those that may have been presented by the social and business changes forced by the pandemic (Cano, 2022).

The UK Music Diversity Report 2022 found that women made up 52.9% of individuals working in the British music industries overall, an increase from 49.1% when the equivalent report was undertaken in 2020 (UK Music, 2020). 2022 was the first time women's participation reflected societal make-up (ONS, 2022) by meeting or exceeding 51% of workers. However, all four diversity reports UK Music have published bi-annually since 2016 have consistently demonstrated that women's participation in the British music industries is heavily weighted towards the younger and junior areas of the industry and the percentage of women in the music industries rapidly decreases with age, seniority, and level of income. This can be seen in Figures 6.1 and 6.2 below, both of which were assembled by using data from four reports published by UK Music (2016, 2018, 2020, 2022).

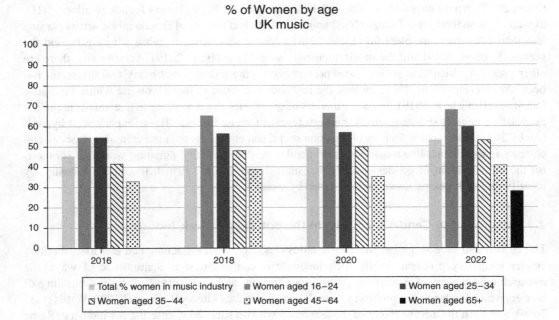

FIGURE 6.1 Data for gender and age in the UK music industry.

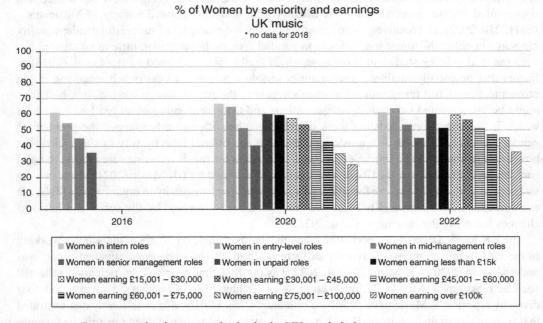

FIGURE 6.2 Data on gender, income and roles in the UK music industry.

However, more encouragingly, year on year, women's participation is increasing in all age groups, levels of seniority and upper salary brackets, although, women do remain under-represented in these areas. My own, ongoing research reflects these findings, with 86.3% of respondents to my 2023 survey stating that gender inequality is an issue in the British live music industry, while 5.9% responded they did not think gender inequality existed in the industry and 7.8% of respondents weren't sure. Additionally, 88% of respondents stated they had either experienced or witnessed gender inequality in their work in the British live music industry. This research along with the previously outlined reports, demonstrates that gender inequality is still a systemic issue within the British live music industry and one that needs urgent attention. "It is about time the music industry took note. The statistics reported prove the music industry is still run by men and rewards them considerably more" (Bain, 2020, p. 1).

There also seems to be very little focus on the issue of gender inequality within the wider music industries as part of its plans to "Build Back Better" than before the pandemic (UK Music, 2021). The 2021 UK Music report: *Music Industry Strategic Recovery Plan*, focused on three key areas: 1. Securing talent; 2. Supporting infrastructure; and 3. Encouraging exports abroad, largely by focusing on additional funding and tax incentives. The report makes no mention of social justice issues within the industry and how they might be improved during this period of re-building. In 2022 and 2023, the Event Production Show (EPS, 2023) took place at Excel, in London following a pandemic-related absence. The annual event is made up of two days of conferences and panels relating to key issues in the live events and music industries. The key themes for the 2022 show were: supply chains and maintaining creativity in relation to Covid-19; social responsibility, relating to legal and safe activities; environmental sustainability; and, safety, security and insurance, in light of major incidents including the Manchester Arena bombing in May 2017, the UEFA Euro 2020 Final crowd disturbance at Wembley Stadium in July 2021 and the Astroworld Festival crowd crush in Houston in November 2021. Although some aspects of social responsibility and sustainability were addressed, the ongoing issue of gender inequality within the industry was certainly not a focus at the Event Production Show 2022, despite being held on International Women's Day. The Event Production Show 2023 contained little to no mention of gender equality and focused heavily on: safety and licensing; marketing and brand experience; and environmental sustainability. Similarly, the International Live Music Conference (ILMC, 2022, 2023), the world's largest conference for music industry professionals, had a focus on industry responses to the Covid-19 pandemic; the impact of Brexit and the war in Ukraine on the music and touring industries; technology and green initiatives for its 2022 conference, and its 2023 edition focused on technology; marketing and social media; environmental sustainability; and health and mental wellness. In other words, it appears, based on the themes outlined, that key music industry organisations and bodies are choosing not to prioritise matters of gender equality as the industry attempts to build back post-pandemic.

3.   Gendered Organisation Theory

Acker's (1990) gendered organisational theory, argues that organisations are inherently male gendered and that in the workplace the term "work force" is stereotyped as male, whereas the term "gender" relates only to women. As such, in discussions and analysis of gender in the workplace, male-ness is often invisible and removed from critical reflection. Men are considered the default and definition by which the "other" is defined. And treats women as the problem. Acker's theory has been developed upon by other gendered organisational theory scholars (Atena and Tiron-Tudor, 2020; Gherardi, 2009; Vandello et al., 2013; and Williams and Muller, 2012) who demonstrated how, in the vast majority of workplaces, men's experiences and behaviours are still seen as the default which women need to meet and the environment created by men is one that women are still expected to move towards. They further note that initiatives that seek to create processes, behaviours, structures, atmospheres and environments that would benefit the needs of all workers, are still too often neglected. Their research relates to additional literature which has focused on women's participation in the workplace and how they can "fit in", as detailed by Gherardi (2013), but not how the understanding of women's behaviours and experiences should inform the overall "worker experience" from a female and male perspective to reveal ideal outcomes for both. Kumra et al. (2014) draw on existing literature, including that by Plowman (2000) and Vakola et al. (2004), when highlighting how traditional organisational change management models perpetuate gender inequality by failing to specifically address gender. Organisations frequently claim they are "gender blind" but by failing to properly address gender, organisations perpetuate the gender inequality problem given that women and men experience differently. Regardless of whether such difference is acknowledged, the ongoing issue is that the majority of organisations, and especially ones within the live music industry, have not explicitly included and ingrained gender equality into their ethos and into the values of their organisational change management and decision-making processes.

Plowman (2000) also argues that by acknowledging and allowing for gendered differences, the barriers faced by women and facilitators for women's access and advancement can be identified and addressed as part of updated organisational aims, strategy and vision, with more overt references to gender inclusivity included within these. Similarly, Platt and Finkel (2018) highlighted that previous attempts at achieving gender equality in the professionally planned events sector have not been ingrained into the ethos of change management. They demonstrated that much more than tokenistic or short-term initiatives are needed if ingrained inequalities are to be successfully confronted in the planned events sector and warned that minor attempts to address these problems may in fact do more to disguise them than to address them.

Rodriguez and Guenther (2022) state that organisation theory remains largely "gender-neutral" but that the default of gender-neutral is usually male. They argue that gender is a basis for inequality in working life and that it needs to be explicitly included within organisational management in order for change to occur and for gender inequality to be improved. Edmond (2019) criticises previous attempts to implement positive change in the area of gender inequality in the creative industries as having a tendency to return to the same types of interventions repeatedly, which have demonstrably not worked. Edmond also highlighted the lack of explicit gender equality in Cultural Policy discourse and how gender equality is most often addressed implicitly as a generalised concept which includes terms such as "diversity", "inclusion", and, or "access for everyone". It is very broadly defined and without clear objectives, especially by way of how they apply to gender. This can be demonstrated in the British music industry, such as Live Nation UK, who claim to promote gender equality, but have a mean gender pay gap of 35%, and where only 29% of promoters are women, do not appear to have a gender equality-specific policy, but their Equal Opportunities policy (Live Nation, 2023) states:

> At Live Nation, we need to be as diverse as the fans and artists that we serve. We're continually striving towards this goal on all fronts to uplift people across race, ethnicity, gender, sexual orientation, disability, and other underrepresented groups. The core of our business is promoting. And we are committed to improving our promotion of diversity within our company and the world at large.
>
> *(Ibid.)*.

Similarly, AEG Presents, also claim to be an inclusive organisation (AEG Europe, 2023) despite a mean gender pay gap of 37.5%:

> AEG is an inclusive organisation where we value everybody's contribution. We are committed to fostering a diverse and inclusive workforce. We believe in equality of opportunity for all and our approach to recruitment and selection is fair, open and based purely on merit. Applications from individuals are encouraged regardless of age, disability, sex, gender reassignment, sexual orientation, pregnancy and maternity, race, religion or belief and marriage and civil partnerships.
>
> *(Ibid.)*

These policies reflect the points made by Edmond (2019), that they are too broad and too vague and in no way explain how such inclusiveness and equality might manifest in the music industry. Nor do they outline any objectives, tools or mechanisms to help achieve this. Furthermore, there are not clear objectives to meet the different needs of the very different characteristics of inclusivity and equality, such as gender, ethnicity, race, disability, age, etc., which each have their own challenges, and barriers to access and advancement within the industry, and as such, their own facilitators to access and advancement to be investigated. To group together such a broad intersection of people, who face, often vastly different challenges and barriers from one another, into one group called "equality" or "diversity" or "inclusion", appears counter-productive at best and lazy at worst. However, such broad and vague policies appear to be commonplace for the music industry organisations who have such policies at all, although many industry organisations appear to have no such policies at all, at least not publicly visible. Additionally, there are a number of important organisations that exist to support and empower women with networking, training, mentoring and safe spaces in the music industry, such as *Network of Women in Events* (NOWIE) and *Women in Live Music* (WILM); and initiatives such as *Girls I Rate* and the *Keychange Project* exist to work towards a 50/50 gender split of festival and concert line-ups. But previous and existing initiatives to advance women's access in the industry have usually been time-limited and operate as projects or groups separate from industry organisations, rather than gender equality initiatives, policies and procedures being embedded into the culture and behaviours of industry organisations. They reflect the criticisms of tokenism made earlier by Platt and Finkel (2018). In adopting such a stance, these scholars collectively support the important point made much earlier by Acker (1990, p. 146), namely that "Gender is not an addition to an ongoing process, conceived as gender neutral. Rather, it is an integral part of processes which cannot be properly understood without analysis of gender."

4. Barriers and Facilitators to Women's Access and Advancement

Although my research is currently ongoing, early results along with existing findings by researchers in relation to barriers that exist for women in the wider workplace, as well as potential facilitators to their access to and advancement within the workplace have allowed me to develop hypothesised barriers for women and facilitators for their advancement within the music industries, which are key to developing strategies for their implementation.

Coleman (2020) identifies barriers to access and advancement for women in the workplace which can be summarised as: a masculine work culture and environment, which facilitates the maintaining of the status quo where male-led networking and men dominating leadership roles is the norm; gender stereotyping, which favours men and disadvantages women as leaders and which assumes domestic, child and caring responsibilities belong to women, not the responsibilities of leadership; direct and indirect discrimination as well as micropolitical behaviours, such as hostility, undermining, less pay, shorter or less secure contracts; the expectation of women to support not lead and expectations to follow the male model of working long, unsocial hours; and additional invisible hurdles for women in the workplace such as the glass ceiling effect (Cotter et al., 2001) which represents an unseen barrier through which women can see senior and better paid roles, but cannot reach them; and the glass cliff effect (Ryan and Haslam, 2005) whereby women are likelier than men to achieve a leadership role during periods of crisis, when the risk of failure is highest.

In relation to barriers to gender equality in the music industries specifically, Edmond (2019) notes that women are represented well as audiences, students and as early career practitioners, but "with each hurdle towards success, recognition and something approximating a sustainable career, their numbers dwindle dramatically and entirely disproportionately" (ibid., p. 75), which is reflected in the findings of UK Music from 2016 to 2022. Edmond then goes on to explain how various reports and policies written by academics, government agencies and professional industry organisations, repeatedly note that the key obstacles they identify remain remarkably consistent and identified the following explanations for the ongoing gender gap as: motherhood and caring responsibilities; a 'Boys' Club' environment; confidence gaps; a lack of women in leadership roles; segregation, with women over-represented in traditionally feminised roles and absent from better paid, more career-defining roles; sexism and misogyny; and hostile working environments which include behaviours such as discrimination, harassment and intimidation. More specific examples of barriers to access and advancement for women in the British live music industry have been provided in my current research and include challenges and barriers relating to: breast feeding; childcare and caring responsibilities along with unsympathetic male colleagues; menopause; on-site sanitation facilities; room-sharing arrangements; touring; the close relationship between ageism and sexism; top-level roles in the industry still being held by predominantly white men with a 'boys' club' mentality; being held to different rules or higher standards than men; women not being taken seriously or respected in their role by men; and, inflexible working arrangements and working environments incompatible with family life.

Once some of the specific barriers that can prevent women from access to and advancement within the music industries are understood, the next step is to identify facilitators to women's access and advancement within the industry, in order to overcome the aforementioned barriers. The key facilitators to women's access and advancement in the workplace, as noted or experienced by the respondents in Coleman's (2020) research, were: networking, which has been helpful to women who gain contacts, emotional and professional support, as well as professional development and potential employment opportunities; mentoring, by women, for women; women as role models; structural changes such as board quotas and flexible arrangements for staff; a supportive workplace culture which supports women by providing and facilitating training and mentoring tailored towards women, an awareness of indirect or invisible discrimination and facilitating unconscious bias training to counter this; and finally, workplace flexibility, with an understanding of work-life balance for both women and men. Additional workplace gender equality facilitators, as outlined by the European Institute for Gender Equality (EIGE, 2016), include: alternative hiring practices and HR policies; effective, dynamic and workable policies for promoting gender equality and equal opportunity of access and advancement; actively encouraging a good work-life balance for all staff; and creating an open-minded culture and environment. In my own research, early results have uncovered the following facilitators to women's access and advancement in the British live music industry: family friendly touring and festival sites; on-site breast feeding and improved sanitation facilities; single-room allocations; the importance of top-level male buy-in; and flexible and remote working arrangements, post-Covid. Additionally, my initial research indicates that of those respondents who have experienced gender equality focused initiatives, a high percentage stated these were beneficial to women by helping to remove some of the barriers they face, such as: women-led mentoring (95%) and women as role models (100%); training focused towards women (100%); diversity training (93%) and unconscious bias training (100%); flexible working (95%); and remote working (96%). Overall, 96.3% of respondents stated that women-focused initiatives have been beneficial to them and to improving gender equality in the British live music industry.

5. Conclusion

Once both the barriers holding women back, and the facilitators to assist their access to and advancement within the industry have been identified, solutions for implementation need to be sought, in order that such facilitators can be effectively applied. As outlined previously, many of the initiatives that currently exist to support and empower women in the British music industries, sit outside of industry organisations, as separate, industry-adjacent organisations, supporting industry workers and pushing for change. And those organisations that do have policies relating to gender equality tend to be all-encompassing, vague policies that claim inclusivity for all, without any clear aims and objectives. However, for effective, long-term change to happen, explicit gender equality measures need to be adopted by all music industry organisations and gender equality facilitators embedded into organisational policies, processes and procedures, rather than being included in vague statements committing to equality and inclusivity, without definition. And where such organisational policies, processes and procedures do not exist, creating them is key to the future success of organisations' ability to change, adapt and strive for true equality. As outlined, previous researchers have demonstrated the key issues enabling gender inequality to persist, are that the majority of organisations, and especially ones within the music industries, have not explicitly included and ingrained gender equality and facilitators related to women's access to and advancement within the industry, into their ethos, the cultural values of their organisational change management and decision-making processes. This article argues, that by identifying clear facilitators to women's access and advancement within the industry, existing change management models and processes can be utilised to effectively and explicitly, implement them into change cultures and behaviours within industry organisations. By doing this it will assist in removing barriers to women and ingrain gender equality into the ethos of organisations and their management processes for long-term, sustainable, improved gender equality in the British live music industry.

The scope of this article and my current research is narrow and focuses only on gender, specifically in the British live music industry, from an interdisciplinary perspective but without exploring intersectionality or other marginalised groups of industry workers. However, the premise of this research has the potential to be developed upon and for further research to explore a similar approach to gender equality, but in the wider creative industries, and indeed, all workplaces. Additionally, by focusing on the area of gender equality and what the specific barriers and facilitators to women's access and advancement look like to women in the music industries, there is the potential to undertake similar research in relation to other inequalities in the industry, or other industries, such as race, ethnicity, age or disability, where the specific needs of different marginalised groups can properly be addressed.

References

Acker, J. (1990). Hierarchies, Jobs, Bodies: A Theory of Gendered Organisations, *Gender & Society*, 4(2), pp. 139–158.

Adams, T. M. & Fuller, D. B. (2006). The Words Have Changed but the Ideology Remains the Same: Misogynistic Lyrics in Rap Music, *Journal of Black Studies*, 36(6), pp. 938–957.

AEG Presents. (2023). Our Culture, *AEG Europe*, 04.07.2023. https://careers.aegeurope.com/our-culture. Accessed: 04.07.2023.

Atena, F. W. & Tiron-Tudor, A. (2020). Gender as a Dimension of Inequality in Accounting Organizations and Developmental HR Strategies, *Administrative Sciences*, 10(1), p. 1.

Awbi, A. (2017). You Are Not Alone: Help for Musicians Suffering Mental Health Issues. *M Magazine*, 24.10.2017. www.prsformusic.com/m-magazine/features/not-alone-help-musicians-suffering-mental-health-issues/. Accessed: 23.06.2021.

Bain, V. (2019). Counting the Music Industry: The Gender Gap. A Study of Gender Inequality in the UK Music Industry, *Counting Music, UK Music*, 15.10.2019. www.ukmusic.org/wp-content/uploads/2020/09/Counting-the-Music-Industry-full-report-2019.pdf Accessed: 25.03.2021.

Bain, V. (2020). The Gender Pay-Gap in Music. *V Bain Consulting*, 08.07.2020. https://vbain.co.uk/the-gender-pay-gap-in-music/. Accessed: 06.06.2021.

Bayton, M. (2003). *Frock Rock: Women Performing Popular Music*. Oxford: Oxford University Press.

Berkers, P., Smeulders, E., & Berghman, M. (2019). Music creators and gender inequality in the Dutch music sector, *Tijdschrift voor Genderstudies*, 22(1), pp. 27–44.

Boot, F. H., Owuor, J., Dinsmore, J., & MacLachlan, M. (2018). Access to assistive technology for people with intellectual disabilities: a systematic review to identify barriers and facilitators, *Journal of Intellectual Disability Research*, 62(10), pp. 900–921.

Bradshaw, M. (2019). Classical Music's industry survey reveals serious shortcomings in mental health provision, *Rhinegold*, 10.10.2019. www.classical-music.uk/resources/article/classical-music-s-mental-health-survey-reveals-serious-shortage-of-provision. Accessed: 23.06.2021.

Brown, M. (2021). UK Musicians Back Call to End Harassment of Women at Live Gigs. *The Guardian*, 13.08.2021. theguardian.com/world/2021/aug/13/uk-musicians-back-call-to-end-harassment-of-women-at-live-gigs. Accessed: 13.08.2021.

Cannizzo, F. & Strong, C. (2020). 'Put some balls on that women': Gendered repertoires of inequality in screen composers' careers, *Gender, Work & Organization*, 27(6), p. 1346–1360.

Cano, M. (2022). Every Covid Has a Silver Lining: Women's Lived Experiences and Potential Feminist Futures beyond the Pandemic, *Women's Studies*, 51(5), pp. 597–618.

Cobb, M. D. & Boettcher, W. A. III. (2007). Ambivalent Sexism and Misogynistic Rap Music: Does Exposure to Eminem Increase Sexism? *Journal of Applied Social Psychology*, 37(12), pp. 3025–3042.

Coleman, M. (2020). Women Leaders in the Workplace: Perceptions of Career Barriers, Facilitators and Change, *Irish Educational Studies*, 39(2), pp. 233–253.

Conor, B., Gill, R., & Taylor, S. (2015). Gender and Creative Labour, *The Sociological Review*, 63(1), pp. 1–22.

Cotter, D. A., Hermsen, J. M., Ovadia, S., & Vanneman, R. (2001). The Glass Ceiling Effect, *Social Forces*, 80(2), pp. 655–681.

Edmond, M. (2019). Gender, Policy and Popular Music in Australia: 'I Think the Main Obstacles Are Men and Older Men'. In: Raine, S., & Strong, C. (Eds.). *Towards Gender Equality in the Music Industry: Education, Practice and Strategies for Change*. New York, Bloomsbury Academic, pp. 73–87.

European Institute for Gender Equality (EIGE). (2016). Institutional Transformation Gender Mainstreaming: A Guide to Organisational Change, *European Institute of Gender Equality*, 04.10.2016. https://eige.europa.eu/gender-mainstreaming/toolkits/gender-institutional-transformation/gender-mainstreaming-guide-organisational-change. Accessed: 23.10.2023.

Event Production Show (EPS). (2022). www.eventproductionshow.co.uk/2022-programme. Accessed: 02.02.2022.

Event Production Show (EPS). (2023). www.eventproductionshow.co.uk/2023-conference-programme. Accessed: 13.03.2023.

Gallop, J. (2021). Just 21% of Acts at the UK's Top Festivals This Year Are Female. *Access All Areas*, 14.07.2021. https://accessaa.co.uk/just-21-of-acts-at-the-uks-top-festivals-this-year-are-female/. Accessed: 19.07.2021.

Gherardi, S. (2009). Feminist Theory and Organization Theory: A Dialogue on New Bases. In: Tsoukas, H. & Knudsen, C. (Eds.). *The Oxford Handbook of Organization Theory*. Oxford, Oxford University Press, pp. 210–236.

Gherardi, S. (2013). Organizations as Symbolic Gendered Orders. In: Kumra, S., Simpson, R., & Burke, R. J. (Eds.). *The Oxford Handbook of Gender in Organizations*. Oxford, Oxford University Press, pp. 76–96.

Gill, R. (2002). Cool, Creative and Egalitarian? Exploring Gender in Project-based New Media Work, *Information and Communication Studies*, 5(1), pp. 70–89.

Henry, C. (2009). Women and the Creative Industries: Exploring the Popular Appeal, *Creative Industries Journal*, 2(2), pp. 143–160.

Hill, R. L., Hesmondhalgh, D., & Megson, M. (2020). Sexual Violence at Live Music Events: Experiences, Responses and Prevention, *International Journal of Cultural Studies*, 23(3), pp. 368–384.

Houston, C. (2021). Rip It Up + Start Again: Why Male-heavy Music Festivals Should be Left to Scrap Among Themselves. *Gigwise*, 18.05.2021. www.gigwise.com/features/3399148/rip-it-up---start-again--why-male-heavy-festivals-should-be-left-to-get-scrap-among-themselves. Accessed: 23.06.2021.

Incorporated Society of Musicians (ISM). (2018). Dignity at Work: A Survey of Discrimination in the Music Sector, *Incorporated Society of Musicians*, April 2018. www.ism.org/images/images/ISM_Dignity-at-work-April-2018.pdf

Incorporated Society of Musicians (ISM). (2022). Dignity at Work 2: Discrimination in the Music Sector, *Incorporated Society of Musicians*, September 2022. www.ism.org/dignity-at-work-2-discrimination-in-the-music-sector. Accessed: 06.10.2022.

International Live Music Conference (ILMC). (2022). https://34.ilmc.com/. Accessed: 28.02.2022.

International Live Music Conference (ILMC). (2023). https://35.ilmc.com/. Accessed: 13.03.2023.

Kumra, S., Simpson, R., & Burke, R. J. (Eds.). (2014). *The Oxford Handbook of Gender in Organizations* (First Edition). Oxford, UK: Oxford University Press.

Lieb, K. J. (2018). *Gender, Branding and the Modern Music Industry: The Social Construction of Female Popular Music Stars*. London, Routledge.

Live Nation Entertainment. (2023). Work With Equality, *Live Nation Entertainment*, 04.07.2023. www.livenation.co.uk/about. Accessed: 04.07.23.

Malt, A. (2021). Moving The Needle Aims to Support Women to Rise Up to Senior Roles in the Music Industry. *CMU – Complete Music Update*, 24.02.2021. https://completemusicupdate.com/article/moving-the-needle-aims-to-support-women-to-rise-up-to-senior-roles-in-the-music-industry/. Accessed: 23.06.2021.

Niethammer, C. (2019). The Equivocal Business Of Women In The Music Industry. *Forbes*, 01.12.2019. www.forbes.com/sites/carmenniethammer/2019/12/01/the-equivocal-business-of-women-in-the-music-industry/. Accessed: 08.05.2021.

Office of National Statistics. (2022). Population and Household Estimates, England and Wales: Census 2021, *ONS*, 28.06.2022. www.ons.gov.uk/peoplepopulationandcommunity/populationandmigration/populationestimates/bulletins/populationandhouseholdestimatesenglandandwales/census2021. Accessed: 28.03.2023.

Oluyede, L., Cochran, A. L., Prunkl, L., Wang, J., Wolfe, M., & McDonald, N. C. (2022). Unpacking Transportation Barriers and Facilitators to Accessing Health Care: Interviews with Care Coordinators, *Transportation Research Interdisciplinary Perspectives*, 13, article.100565.

Platt, L. & Finkel, R. (2018). Editorial: Special Issue on Equality and Diversity in the Professional Planned Events Sector, *Journal of Policy Research in Tourism, Leisure and Events*, 10(2), pp. 113–116.

Plowman, P. (2000). Organisational Change from Two Perspectives: Gender and Organisational Development, *Development in Practice*, 10(2), pp. 189–203.

Raine, S. & Strong, C. (2018). Gender Politics in the Music Industry, *IASPM Journal*, 8(1), pp. 2–8.

Raine, S. & Strong, C. (Eds.). (2021). *Towards Gender Equality in the Music Industry: Education, Practice and Strategies for Change*. New York, Bloomsbury Academic.

Reimer, S. (2016). It's Just a Very Male Industry: Gender and Work in UK Design Agencies, *Gender, Place and Culture*, 23(7), pp. 1033–1046.

Rodriguez, J. K. & Guenther, E. A. (2022). Gendered Organization Theory. In: Hitt, M. A. (ed.). *Oxford Research Encyclopedia of Business and Management*, Oxford Research Encyclopaedias, Oxford University Press. pp. 1–35.

Ryan, M. K. & Haslam, S. A. (2005). The Glass Cliff: Evidence that Women are Over-Represented in Precarious Leadership Positions, *British Journal of Management*, 16(2), pp. 81–90.

Samuelson, K. (2019). Women Grossly Underrepresented in the Music Industry. *Northwestern University*, 16.09.2019. https://news.northwestern.edu/stories/2019/09/women-are-underrepresented-in-the-music-industry/. Accessed: 29.04.2021.

Scenestr. (2021). Women Call for an End to Toxic Culture in the Australian Music Industry. *Scenestr*, 18.05.2021. https://scenestr.com.au/music/women-call-for-an-end-to-toxic-culture-in-the-australian-music-industry-20210518). Accessed: 23.06.2021.

Stassen, M. (2019). Revealed: What Major Labels Are Paying Women Compared to Men in the UK. *Music Business Worldwide*, 04.04.2019. www.musicbusinessworldwide.com/revealed-what-major-labels-are-paying-women-compared-to-men-in-the-uk/. Accessed: 14.05.2021.

Strong, C. & Rush, E. (2018). Musical Genius and/or Nasty Piece of Work? Dealing with Violence and Sexual Assault in Accounts of Popular Music's Past, *Continuum*, 32(5), pp. 569–580.

Tabbush, C. & Friedman, E. J. (2020). Feminist Activism Confronts COVID-19, *Feminist Studies*, 46(3), pp. 629–638.

Taylor, M. & O'Brien, D. (2017). 'Culture Is a Meritocracy': Why Creative Workers' Attitudes May Reinforce Social Inequality, *Sociological Research Online*, 22(4), pp. 27–47.

UK Music. (2016). UK Music Diversity Report 2016. *UK Music*, 09.01.2017. www.ukmusic.org/equality-diversity/past-projects-2/uk-music-diversity-report-2016/. Accessed: 05.05.2023.

UK Music. (2018). UK Music Diversity Report 2018. *UK Music*, 01.11.2018. www.ukmusic.org/wp-content/uploads/2020/08/UK_Music_Diversity_Report_2018.pdf. Accessed: 05.05.2023.

UK Music. (2020). UK Music Diversity Report 2020. *UK Music*, 02.11.2020. www.ukmusic.org/wp-content/uploads/2020/11/UK_Music_Diversity_Report_2020.pdf. Accessed: 23.11.2022.

UK Music. (2021). Music Industry Strategic Recovery Plan. *UK Music*, 24.09.2021. www.ukmusic.org/news/uk-music-unveils-blueprint-to-boost-growth-and-jobs-in-covid-hit-music-industry/. Accessed: 28.11.2021.

UK Music. (2022). UK Music Diversity Report 2022. *UK Music*, 22.11.2022. www.ukmusic.org/news/uk-music-reveals-2022-workforce-diversity-survey-and-unveils-new-plan-to-boost-diversity-and-inclusion/. Accessed: 23.11.2022.

Vakola, M., Tsaousis, I., & Nikolaou, I. (2004). The Role of Emotional Intelligence and Personality Variables on Attitudes toward Organisational Change, *Journal of Managerial Psychology*, 19(2), pp. 88–110.

Vandello, J. A., Hettinger, V. E., Bosson, J. K., & Siddiqi, J. (2013). When Equal Isn't Really Equal: The Masculine Dilemma of Seeking Work Flexibility, *Journal of Social Issues*, 69(2), pp. 303–321.

Wang, Y. & Horvát, E. A. (2019). Gender Differences in the Global Music Industry: Evidence from MusicBrainz and The Echo Nest, *Proceedings of the International AAAI Conference on Web and Social Media*, 13(1), pp. 517–526.

Wattis, L., Yerkes, M., Lloyd, S., Hernandez, M., Dawson, L., & Standing, K. (2006). Combining Work and Family Life: Removing the Barriers to Women's Progression; Experiences from the UK and the Netherlands, *School of Social Science*, Liverpool John Moores University. www.researchgate.net/profile/Louise-Wattis/publication/242149927_Combining_Work_and_Family_Life_Removing_the_Barriers_to_Women%27s_Progression_Experiences_from_the_UK_and_the_Netherlands/links/568d165a08aef5c20c147bf4/Combining-Work-and-Family-Life-Removing-the-Barriers-to-Womens-Progression-Experiences-from-the-UK-and-the-Netherlands.pdf

Weitzer, R. & Kubrin, C. E. (2009). Misogyny in Rap Music: A Content Analysis of Prevalence and Meanings, *Men and Masculinities*, 12(1), pp. 3–29.

Williams, C., Muller, C., & Kilanski, K. (2012). Gendered Organizations in the New Economy, *Gender & Society*, 26(4), pp. 549–573.

Williamson, M. (2010). Female Celebrities and the Media: The Gendered Denigration of the 'Ordinary' Celebrity, *Celebrity Studies*, 1(1), pp. 118–120.

7

MODES OF ENGAGEMENT WITH CLASSICAL MUSIC

Digital Formats

Clara Colotti

1. Introduction

In response to government legislation during the onset of the COVID-19 pandemic, orchestras were obliged to cancel their live performances (Prime Minister's Office, 2020; Presidenza del Consiglio dei Ministri, 2020). As a consequence, numerous orchestras initiated the delivery of events online as a necessity to establish interaction with audiences. This resulted in an eruption of performances delivered online throughout the pandemic. However, certain orchestras have been engaged in concert streaming since the latter part of the 2000s. The Detroit Symphony Orchestra launched their digital platform in 2011 to reach a larger number of people, and become a more inclusive and culturally relevant institution (Chucherdwatanasak, 2020). The Berliner Philharmoniker launched their Digital Concert Hall in 2008 with the intention of expanding the orchestra's audience reach (Kavanagh, 2018).

As the 2021–2022 season unfolded, orchestras worldwide gradually resumed their live performances. Some orchestras chose to continue offering concerts online alongside live events, while others predominantly reverted to live performances. The aim of this chapter is twofold:

- investigate the advantages and disadvantages experienced by orchestras when presenting streamed events in video formats
- analyse the extent to which these formats can support orchestras in attracting new audience segments.

To ensure the delineation of a coherent analytical framework, this research focuses exclusively on orchestras of international prominence, each distinguished by their interpretation and execution of classical music repertoire.

Section 2 explores classical music audiences. Section 2.1 gives an overview of the demographic characteristics of classical music audiences; Section 2.2 investigates what motivates individuals to attend classical music concerts; Section 2.3 illustrates varying degrees of familiarity and involvement that audiences could exhibit in relation to classical music attendance. Afterwards, Section 3 analyses the current landscape of classical music attendance. In the wake of the diminished attendance in classical music, as outlined in Section 3, Section 4 delves into the underlying reasons prompting orchestras to establish connections with new audience segments. Finally, Section 5

DOI: 10.4324/9781003396550-7

explores the potential of digital formats in facilitating outreach to new audiences. Section 5.1 examines advantages and disadvantages for orchestras to present events in digital formats. Section 5.2 draws on empirical research and its main objective is to examine the viewpoints held by the surveyed orchestras concerning the influence of video-format events on audience engagement.

2. Characterising the Classical Music Audience

Understanding the classical music audience holds significance for various stakeholders within the classical music sector. By understanding what resonates with the audience, classical music institutions and performers can tailor their programmes, repertoire, and interpretations accordingly (Heilbrun and Gray, 2001). In addition, understanding the demographics of the classical music audience helps in developing targeted marketing strategies and outreach programmes (Kolb, 2005). Furthermore, understanding the preferences and responses of the audience can influence artistic decision making. Artists and conductors may take into account audience feedback to shape future programming and interpretations, striking a balance between artistic integrity and audience appeal (Turrini et al., 2008). Moreover, classical music has a rich cultural heritage, and understanding the audience helps in preserving and promoting this art form (Brown and Novak, 2007). By catering to contemporary tastes while staying true to tradition, classical music remains relevant in a changing society (ibid.). Finally, for orchestras, relying on funding and sponsorships, knowing the demographics and interests of the audience can attract potential sponsors who share a passion for classical music, ensuring financial stability (Turrini et al., 2008).

The purpose of the next sections is to examine the composition of classical music audiences. The key elements explored are the demographics and socioeconomic factors, as well as the motivations that influence audience participation.

2.1. Overview of the Demographic Characteristics of Classical Music Audiences

A report released by The Audience Agency shows that the demographics of classical music audiences in the United Kingdom are significantly skewed towards middle and older age cohorts. Specifically, individuals aged 41 to 60 constitute approximately 4 per cent of the audience, while those aged over 61 account for approximately 37 per cent. Conversely, the younger age group of individuals under 31 comprises merely 7 per cent of the audience, reflecting a comparatively lower level of engagement among this segment (Bradley, 2017). The German specialist magazine *Concerti* published a study on classical music attendance in 2016. Their findings assert that 31.5 per cent of classical music concertgoers in Germany are 60 years or older (*Concerti*, 2016). A survey of the performing arts in Spain indicates that 40 per cent of the audience is 45 or over (Ministerio de Cultura y Deporte, 2021, p.29). Research on audiences in the United States reports that the median age of classical music attendees is 53 (Gaylin, 2016, p.212).

The data included in the report by The Audience Agency, *Concerti*, the Ministerio de Cultura y Deporte in Spain and Gayilin (2016) are reflected in O'Sullivan (2009), Pitts et al. (2013) and Crawford et al. (2014) studies on classical music audiences. O'Sullivan's (2009) research highlights a recurring theme that emerges as a significant subject of concern throughout the study: the absence of a younger or more diversified audience attending live classical music performances. Pitts et al.'s (2013) audience survey shows a persistent concern over an ever-increasing decline and ageing of attendees of the City of Birmingham Symphony Orchestra's season (Pitts et al., 2013). Research by Crawford et al. (2014) shows that the typical demographic of concert attendees in both the UK and the US are predominantly "older and white" (Crawford et al., 2014, p.485).

In terms of educational background, the Ministerio de Cultura y Deporte in Spain (2021) and Gans (1998) show that classical music audiences have a high degree of education. According to the Spanish report, 20 per cent of concert goers have a university degree, 8 per cent advanced vocational training qualifications, 16 per cent other higher education degrees (Ministerio de Cultura y Deporte, 2021, p.29). Gans (1999) classifies culture in four different strata: high, middle-class, lower-middle-class and working-class cultures. The author argues that the audience of high culture (for example classical music) is formed by individuals with high levels of education, necessary to interpret its meaning.

2.2. *Motivations that Drive Individuals to Attend Classical Music Concerts*

Individuals can be motivated by a combination of factors to attend classical music performances. First, attendees may be drawn to classical music concerts because they have a genuine appreciation for the artistry and cultural significance of classical music (Thompson, 2007). Research by Dobson (2010) shows that classical music has the power to evoke a wide range of emotions; people may be drawn to attend performances to experience these emotions in a live setting. Caldwell and Woodside (2003) argue that cultural capital may play an impartial motivational role: some individuals attend concerts to engage intellectually with the music and appreciate the technical skills of the musicians. O'Sullivan (2009) observes that while there is a personal connection between concert attendees and the music, attendance of live performances is a social experience. The author investigates patterns of shared consumption concerning the audience experience with the performing arts. His study shows that the act of being physically present at a concert venue alongside other attendees, who share similar intentions, contributes to the creation of a communal experience (ibid.). A study by Roose (2008) analyses the relevance of the aesthetic experience for the audience, including the ambiance of a concert hall, and the lighting. His study demonstrates that the overall atmosphere contributes to the aesthetic experience, making the concert not just about the music but also about the entire sensory journey. Moreover, DeNora (1997) argues that attending a classical music concert provides an opportunity to escape the stresses of everyday life and has the ability to move people emotionally. Finally, concert halls where classical music is performed are purpose-built spaces with specific architecture and sophisticated acoustics (Barron, 2010). Participating in a live classical music event offers an opportunity to engage with an event transpiring within an environment where the auditory experience evokes excitement and emotional resonance (ibid.). To conclude, reasons for classical music performances attendance are not static and may evolve over time.

2.3. *Knowledge of and Engagement with Live Classical Music*

The previous section has shown how audiences can attend classical music performances for various reasons. This section demonstrates that audiences may hold different levels of knowledge and engagement. Within the context of classical music audiences, a prominent category comprises enthusiasts, denoting individuals who actively and extensively engage with the genre, demonstrating profound expertise and familiarity with the classical music repertoire; Roose (2008) defines this category as core audience, who fulfil the role of gate-keepers and are usually professionally involved in the arts. A second category includes casual listeners: this group enjoys classical music but may not have an in-depth knowledge of the genre. They attend concerts occasionally or listen to classical music in a more relaxed and leisurely manner. Casual listeners may not be familiar with the technical aspects of classical music, but they appreciate the emotional and aesthetic qualities.

This group may include listeners who attend concerts with programmes with film music or video game music. Other categories include first-time concertgoers, seasonal attendees, cultural explorers, and supporters of the arts. All these categories are not mutually exclusive: individuals may fall into more than one group depending on their level of interest and engagement with classical music. Additionally, their preferences and interests may change over time, so they might move from one category to another as they deepen their knowledge of and connection to the genre. Amidst the various discernible categories of classical music audiences, this paper focuses its examination on a specific subset, the aforementioned core audiences, who represent the first and most involved group within the spectrum.

3. Current Landscape of Classical Music Attendance: Trends and Patterns

Media outlets have reported the "death of classical music" for decades (see for example Albright, 2016). Researchers have addressed the issue of the ageing and dwindling of classical music audiences attending orchestral performances in the United Kingdom (O'Sullivan, 2009; Pitts et al., 2013) and Germany (see for example Stauber, 2022); young generations losing contact with tradition and classical music (McEwen, 2015); the decline of music education (Büscher, 2023); music criticism decreasing both in quantity and in quality (Vroon, 2014).

In this particular context, scholars hold divergent viewpoints and perspectives. The debate of a "dying" classical music culture seems to have grown in range and intensity, which Borstlap (2012) interprets as a reflection of a general feeling of malaise in the context of European society, encompassing its culture, cultural identity, values, and prospects in an increasingly globalised world. Borstlap (2012) argues that overarching trends in Western society, which the author regards as characterised by psychological distress and decadence, have resulted in the marginalisation of classical music within public domains.

Other scholars tackle the matter of technology and its repercussions on audience behaviours. In a seminal scholarly work, Hutchby (2001) argues that technologies can be comprehended as artefacts that are subject to mutual influence. According to the author, technologies play a formative role in shaping human behaviours and practices. Conversely, when individuals engage with technologies, these technologies are in turn moulded by their usage. This conceptualisation shows the mutuality between audiences and technologies. In alignment with this perspective, De Ridder (2016) and Naveed et al. (2017) raise a similar point regarding influence of digital media on audiences. According to their studies, consumers have transitioned beyond the role of passive listeners; they now seek an extensive array of choices and demonstrate a willingness to engage actively, integrate themselves, and collaboratively generate value. While De Ridder and Naveed et al. do not take account of the distinction among different audience segments, these shifts are relevant to every musical context. In the realm of classical music, where audiences conventionally assume a passive role, the evolving trend of engaging audiences in active participation throughout a performance signifies a shift. Furthermore, the widespread digitalisation of cultural consumption is believed to yield a comparable influence on the taste of consumers, which can be attributed to the unprecedented variety of cultural content accessible online, often at minimal or no expenses (Webster, 2020; Walmsley, 2016). In this scenario, Borstlap (2012) argues that popular culture, characterised by its immediate entertainment value, has gained dominance.

Pierre Bourdieu's (1984) perspective on taste underlines the influence of societal aspects on the taste of an individual and makes a hierarchical distinction between cultural systems. Peterson and Simkus (1992), among others, argue that distinct boundaries between societal classes are no longer as distinct as they were when Bourdieu formulated his theory. According to Bourdieu, on one side of the spectrum sits "highbrow culture", a term he uses to indicate tastes and cultural practices that have historically been consumed by the upper-middle and upper classes. For Bourdieu, classical music belongs to highbrow culture. Respectively, the term "lowbrow culture" applies to tastes and cultural practices that have been used by the poor and working classes. Bourdieu's theory has been empirically tested by Peterson and Kern (1996) in a US context: as a result of their survey, the authors formulate their theory of the *cultural omnivore*. Omnivorousness is defined by consumption patterns which are open to a variety of cultural activities, replacing patterns of highbrow and lowbrow consumption. Webster (2020) drives similar conclusions regarding the argument of digital and online technology leading to shifts in cultural consumption patterns, and ultimately to an *omnivorisation* of culture tastes.

Within the framework of declining classical music attendance, what prompts considering the perpetuation of classical music? Studies demonstrate that preserving the heritage of classical music holds significant importance for several reasons. Firstly, Ramnarine (2011) contends that orchestras assume significance within the context of civil society by engaging with matters pertaining to community dynamics, diversity, poverty, ecological concerns, and the principles of sustainability. José Antonio Abreu's vision of "art" being available to marginalised communities provides an example from the Venezuelan context. Implemented in 1975, the extensive network of youth orchestras and music education centres, widely known as *El Sistema*, strives to provide free classical music education, thereby fostering human prospects and advancement among underprivileged children (Tunstall, 2013).

Secondly, Taruskin (1997) argues that classical music stands as a testament to the cultural and artistic achievements of past generations, reflecting the historical context, societal values, and artistic sensibilities of their eras. The preservation of classical music ensures the continuity of a rich cultural legacy, fostering a sense of historical continuity and connection between present and past. The relevance of classical music within contemporary non-Western societies, particularly in Asia, despite the origin of this genre within a distinct historical epoch on the European continent, implies that the implications of classical music culture's decline might extend beyond the preoccupations of a "minor elitist faction" (Botstein, 2020).

Furthermore, classical music encompasses a diverse range of styles, genres, and compositional techniques that have been refined over centuries. This reservoir of musical knowledge provides invaluable resources for contemporary composers, performers, and scholars, contributing to the evolution and innovation of musical expression (Botstein, 2020). A notable episode is the conversation between composer Erich Wolfgang Korngold and a friend during a social event in Hollywood in the late 1940s. While they happened to wander in the music room of a fellow composer, the shelves held a comprehensive collection of scores authored by revered canonical composers. In response to his friend's astonishment, Korngold remarked, "now you see where he gets all his material" (ibid.).

Moreover, according to American musicologist and composer Lawrence Kramer (2007), music often conveys profound emotional and intellectual depth, offering a means of introspection, contemplation, and transcendence for both performers and audiences. Its preservation thus perpetuates an avenue for emotional exploration and personal growth.

Lastly, work by Orchestra Live, a project supported by the Arts Council England, to illustrate one, demonstrates that classical music can be beneficial to health and wellbeing (Arts Council England, 2023).

4. Exploring the Rationale for Orchestras to Engage New Audiences

In response to the decline in classical music attendance addressed in the previous section, the construct delineated by McCartney and Jinnett (2001) for cultivating engagement in the arts argues that arts organisations can build participation in two ways: (1) they can diversify participation by appealing to demographics beyond their existing base; (2) they can deepen engagement depth by increasing the degree of involvement among extant participants. In the context of this conceptual framework, this paper discusses whether orchestras should direct their efforts towards extending their demographic outreach, and subsequently, examines the rationales underpinning the desirability or potential drawbacks of such a pursuit.

Examples from classical music contexts exemplify the feasibility of expanding the reach of classical music to encompass a broader audience spectrum. Barna (2020) critically examines the phenomenon that transpired in the 1990s, wherein music traditionally associated with high culture, such as classical music and opera, attained widespread appeal. This is exemplified by the ascension of The Three Tenors to superstar status: the broadcast of their concert at the Caracalla Baths, in Rome, was broadcast live to 800 million people (Kozinn, 1994). Dromey (2018) investigates the establishment of Classic FM in the United Kingdom as a radio platform dedicated to popularising classical music. Statistics published by Radio Joint Audience Research (RAJAR, 2022) released in February 2023 show that 5 million people listen to Classic FM every week.

Furthermore, my professional engagement within the industry has provided me with discerning insights. Notably, I have observed that artistic directors, marketers, artists affiliated with classical music institutions adopt a fresh approach to programming, and customise promotional campaigns in order to attract new audience segments. As an illustration, orchestras have started integrating different forms of digital technology to create visuals with their live performances in order to reach out to new audience segments (Colotti, 2021). This approach is substantiated by Dobson's (2010) findings, which underscore that new and younger audiences exhibit an interest to the visual component in combination with live classical music performances, thereby evincing heightened receptivity to new concert formats that integrate digital technology. Furthermore, primary research conducted with a sample of international orchestras shows that the artistic planning teams integrate unusual concert formats into their concert seasons. Two participants reported that they include film music in their programming in order to reach new audiences. Another participant replied that they present crossover concerts. A further participant said that they present concerts in shorter formats.

But what motivates orchestras to undertake endeavours aimed at engaging with previously untapped audience segments?

Research shows that the rationale for orchestras to expand their audience reach is multi-faceted and encompasses several fields. First, motivations may stem from the desire to remain financially sustainable: expanding the audience base can lead to increased ticket sales, subscriptions, and donations (McCarthy and Jinnett, 2001). These additional streams of revenue may help orchestras maintain financial stability and support their operations, including financial support for musicians, and staff, and supporting production costs.

A research study by the Royal Philharmonic Orchestra (2023) about patterns of involvement with orchestral music argues that "The future of orchestral music depends on everyone having access to it" (p. 19). Their study encompasses a sample comprising over 700 individuals from black, Asian, and minority ethnic backgrounds, and reveals that orchestral music held significant prominence within their lives (ibid.). Thereby, establishing connections with the most extensive and diverse audience possible is essential to nurture future audiences.

Nonetheless, in the scholarly discourse on this matter, American music critic Donald Richard Vroon (2014) posits an opposing viewpoint. Vroon argues that attempts to enhance accessibility of classical music to diverse audiences might inadvertently result in a lowering of the genre's artistic standards. Furthermore, the music critic argues that with such conduct, marketers tend to prioritise mass appeal over the authentic experiential aspect of live classical music performances.

5. The Potential of Digital Formats for Orchestras in Facilitating Outreach to New Audiences

The start of the COVID-19 pandemic provided a new context for orchestras to try and reach out to audiences, existing and new. Drawing on literature reviews and primary research conducted on a sample of ten international orchestras during the pandemic, the purpose of this section is twofold: Section 5.1. explores the advantages and disadvantages for orchestras of presenting events in digital formats; Section 5.2. analyses to what extent digital formats may support orchestras to engage with new audience segments.

Considering this section examines events in digital formats, terminology deserves some attention. Throughout this research, the term "digital formats" is used to indicate the transmission of concerts in video formats on the internet. It is beyond the scope of this chapter to delve into technical details about internet technologies (for an introduction on streaming and digital media technologies, *Streaming and Digital Media* by Dan Rayburn published in 2005 is a place to start). Relevant to the purposes of this section is the distinction between static and dynamic content on the internet (Rayburn, 2005), as these two concepts illustrate how orchestras employ digital formats to deliver their performances. A video file of a concert available to watch online at any time through an internet connection is an example of static content. Dynamic content includes streaming media, for example the broadcast of a live concert over the internet as it happens. Occasionally, dynamic content is captured and then retained as static content. One example is content which is streamed on YouTube as a live stream event, yet the content has been pre-filmed.

5.1. Challenges and Affordances of Orchestrating Digital Format Events

Szabo and Szedmák (2020) argue that digital formats can attract a public that does not have the ability to attend live performances for different reasons, including remote location, schedule, or other commitments. Furthermore, Webster (2020) posits that music streaming platforms facilitate access to musical content that individuals may have otherwise felt barred from engaging with. Studies by Chucherdwatanasak (2020), Cooper (2016) and Trainer (2015) reveal that digital formats allow performers to share with audiences extra content that is not presented during live events. This allows for a more personalised level of engagement among performers and their audiences (Chucherdwatanasak, 2020), tackling the issue of classical music performances perceived as distant and not engaging by certain audience segments.

Conversely, research by Trainer (2015) determines that the costs to professionally produce content delivered in digital formats are prohibitive. Furthermore, Varney and Fensham (2000) point out that techniques employed during concert filming to reproduce a live performance in a video format may lead to a relinquishment of the viewer's autonomy in determining the focus of their attention. Lastly, studies by Baym and Boyd (2012), and Airoldi (2021) demonstrate that digital technology and events online may accentuate divisions that already exist between different groups and audiences, due to unequal access to an internet connection. Crawford et al. (2014) show that specific segments of the audience exhibit resistance to technology, casting doubt on the feasibility of providing access to events through digital formats.

5.2. Primary Research Findings: Digital Formats Enabling Orchestras to Engage with New Audience Segments

While Section 3 delineated a notable decline in the attendance of classical music performances, this section endeavours to evaluate the potential utilisation of digital formats by orchestras as a means of fostering connections with audience segments new to classical music. The researcher is not aware of any published works that have undertaken an assessment of the prospect that digital formats might offer orchestras in extending their reach to untapped audiences.

This section draws on empirical research gathered directly from orchestras. Overall, the aim of the empirical research is to provide responses to the two points raised in Section 1:

- investigate the advantages and disadvantages experienced by orchestras when presenting events in video formats
- analyse the extent to which these formats can support orchestras in attracting new audience segments.

Methodology

In order to answer the main questions and gain an insight into audience engagement with events delivered in digital formats, data were collected through questionnaires. The choice fell on questionnaires as they have been used as the main methods of data collection in qualitative research on audience engagement in the context of classical music performances (Crawford et al., 2014). The questionnaire is structured, comprising distinct categories of questions: qualitative and quantitative; open and closed; behavioural; opinion and fact; general.

Participants for this questionnaire were selected with purposive sampling (Schutt, 2006). Inclusion criteria comprise of orchestras with established notability that have delivered concerts in digital formats during the pandemic; and orchestras that have continued to offer digital concerts after concert halls reopened. In addition, the criteria of selecting orchestras participating in this study considered their geographic location: orchestras from different continents were contacted for this research to ensure a comprehensive global representation. Forty orchestras were contacted to participate in the main study. Ten participants replied to the questionnaire. These participants hold different roles at their respective institutions, including Marketing Manager, Executive Producer, Development Director, Head of Communications, Digital Media Manager, Head of Audience Development, and Business Analyst. Participants were recruited by email, taking advantage of the instant communication emails allow (Dillman et al., 2014). Questionnaires were administered via email by the researcher and self-completed by the participants, which helps eliminate a significant potential source of bias within the responses (Brace, 2013).

To minimise biased interpretations that could have threatened the reliability of the results, the researcher adopted a reflexivity approach throughout the study (Mackieson et al., 2018). This means that the process of framing the research questions, the selection of research methods, recruitment of participants were constructed through a constant reflection on how the assumptions of the researcher have an impact on the context analysed.

Regarding positionality, conducting research involving participants who have collaborated with the researcher in devising promotional campaigns for digital performances facilitated the identification of orchestras that have sustained involvement with digital endeavours and, to a certain extent, gauged their level of achievement (the researcher has access to the metrics of their advertising campaigns).

In this study, thematic content analysis was coupled with an inductive analysis, a choice aligned with Psathas' (1973) assertion that inductive analysis is the most widely used option for understanding qualitative data. The questionnaire was subsequently desegregated into fragments, with each question forming an individual fragment, as proposed by Watts (2014). Subsequently, key phrases and words were extracted from the responses. The responses were annotated, broken down into separate units of meaning. During this process, known as open coding (Goulding, 2005), data are conceptualised and labelled. The purpose of this step is data analysis, categorisation and comparison, in order to trace a path from the descriptive to the interpretative (ibid.). Finally, key phrases and words were regrouped under a set of thematic headings. In accordance with thematic analysis (Braun and Clarke, 2006), a list of common themes was extracted from the responses and thematic analysis was employed to arrange the data into themes. Thematic analysis examines the data without engaging pre-existing themes and it can therefore be adjusted to any research that relies on the input of participants (Denscombe, 2017), which is the case for this research.

Results

What follows is an examination of the key findings obtained from the main questionnaire. The identified themes from the questionnaire responses directly align with the study's objectives, encompassing an exploration of the benefits and drawbacks perceived by orchestras when offering streamed video events. Additionally, the analysis delves into the potential of these formats to engage new audiences. The qualitative assessment yields themes centred around both audience demographics and the viewing experience. Qualitative responses have been grouped into five main themes: (a) Innovative concert formats; (b) Young demographics; (c) Enhancing accessibility; (d) Visual aspects; (e) Social and communal dimensions of concert attendance.

In theme (a) most participants refer to the practice wherein orchestras present their live and digital performances in varying formats, strategically designed to engage with distinct segments of the audience. One participant mentions shorter concerts in their programming to appeal to younger audiences. Another participant indicates crossover concerts with the orchestra to attract new audiences. An additional participant makes reference to interdisciplinary presentations that amalgamate musical expressions with choreography, theatrical elements, and multimedia components. One of the participants contends that the choice of concert venue significantly influences the demographics of the audience attending the performance.

In relation to theme (b), all participants seem to agree that younger audience demographics are more engaged with digital formats. Illustrating this point, one participant asserts that the audience for their digital concerts differs significantly from their in-person concert attendees, noting that digital viewers tend to be younger. Likewise, a second participant affirms that their digital concert hall proves appealing to their younger audience segments as well. According to a further participant, video serves as a potent medium within this demographic.

Theme (c) illustrates the manner in which orchestras, through digital formats, enhance their accessibility across diverse dimensions, encompassing geographical, social, financial, and temporal considerations. First, all participants argue that online concerts facilitated the engagement of audiences situated across various geographical regions. One of the participants points out that audiences who have engaged with their digital concerts are more widely dispersed geographically. A different participant indicates that digital formats enable the orchestra to reach new local audiences in the region, particularly individuals facing constraints in attending performances at the concert hall. Second, one participant raises the issue of social causes affecting audience attendance of live performances and how online concerts help orchestras to remove social barriers. Moreover, the same participant trusts that online concerts have the potential to transform classical music into a form of entertainment accessible to a wider audience. In a similar way, another participant asserts that digital concerts prove valuable in disseminating and increasing access to culture, in particular when events are free to view. The free aspect of digital concerts raises the issue of financial aspects. Some participants indicate that the cost of a performance has an impact on what audience demographics attend an event. One of the participant reveals that in order to appeal to younger segments of the audience, the orchestra offers tickets at reduced price for specific categories, for example students and those under 30. This echoes the strategy adopted by a different participant, who offer free or discounted tickets to students. Lastly, one participant argues that concerts online offer audiences the option to choose when to watch an event with no time constraint and without concerns related to public transport.

In reference to theme (d), all participants seem to agree that online concert formats have an impact on the visual experience for the audience. One participant argues that digital formats allow the spectator to view musicians at closer range in comparison to their live equivalent. But a different participant warns that in the context of online concerts the filming director is expected to have familiarity with the score, and the team should be equipped with appropriate tools in order to provide quality video products. As per one participant, this concern can be mitigated by crafting concerts that incorporate visual artistic decisions tailored specifically for online performances. The incorporation of visual elements into performances seems to enhance the experience for audiences during in-person contexts. However, another participant contends that reproducing these visuals in an online concert proves challenging: staged concerts or multidisciplinary performances involving video and light design are more effective in a direct interaction setting and pose difficulties in terms of filming and online broadcasting.

In regard to theme (e), all participants seem to agree that there is a social dimension to attending live performances. While one participant argues that orchestras can engage with audiences online, a different participant contends that the absence of personal connections and interactions with patrons is a notable drawback when activities are conducted virtually. According to the latter participant, replicating such dynamics in virtual contexts has proven challenging.

6. Discussion

This section addresses the two points presented in Section 1. Subsequently, it engages in a discussion of the prominent themes derived from the questionnaire to address these concerns.

The outcomes derived from the administered questionnaire indicate that digital formats are an effective tool for orchestras to engage with audiences. Firstly, these formats help orchestras to connect with their community. Moreover, the participants of this study seem to agree that digital formats increase their visibility abroad. This reflects the point made by Follansbee (2006), who argues that streaming services have a global reach and increase opportunities to engage with audiences. In addition, the responses to the questionnaires highlight how orchestras may overcome social obstacles through digital formats. Social class and exclusion in live classical music events have been documented (Crawford et al., 2014). Younger audiences often find traditional classical concerts intimidating due to their setup (ibid.). However, watching a concert from home may be less intimidating. Conversely, attending live performances is a social experience and there is a sense of community and shared experience (O'Sullivan, 2009). In this context, one of the participants suggests that watching an online concert from home does not constitute a social activity. Finally, another participant raises the point of digital performances presenting a significant supplementary expense on orchestras. This finding is consistent with those of Cooper (2016), Szabo and Szedmák (2020), and Trainer (2015), who assert that professionally produced content delivered in digital formats has prohibitive costs. In addition, new technology presents additional costs related to the extra time and energy necessary to learn how to use it efficiently (Tepper and Hargittai, 2009). Nevertheless, from the perspective of the audience, online events tend to be financially more accessible. In accordance with comments by one of the participants, digital concerts may increase access to culture, in particular when events are free to view. This echoes findings by Berger et al. (2015), who contend that online audiences tend to gravitate towards free content, thereby expanding the potential to reach novel viewers. Nevertheless, crafting a sustainable business model for online content remains intricate and an ongoing challenge (Rußell et al., 2020).

The second question seeks to determine whether orchestras can reach new audience demographics with the support of digital formats. Peukert (2019) argues that audio streaming platforms constitute the primary avenue through which younger generations engage with music. The findings from the administered questionnaire demonstrate similar results. All participants convey that, in comparison to audiences attending live performances, younger demographics engage with online concerts. These results align with the report by the British Phonographic Industry, Deezer and the Royal Philharmonic Orchestra (2020), indicating that over two-thirds of Deezer subscribers who listen to classical music are aged below 35.

One participant mentions that precisely focused events within the concert hall setting seem to help orchestras to attract new and younger audience demographics to live performances. However, a minority of respondents refer to online concerts as an efficient means to access new audience segments. For instance, one participant asserts that "video works well because is a strong media in younger audiences". Nevertheless, these data should be interpreted with caution, as not only young audience demographics have become accustomed to viewing music performances on screens; the popularisation of television has also played a role in predisposing the public to visual communication (Greckel, 1992).

7. Conclusions

This chapter offers a critical, but by no means exhaustive, overview of relevant theoretical approaches to classical music attendance, and complements them with an empirical approach that investigates the impact of digital formats on patterns of attendance. The purpose of the study was to determine whether overall, it is convenient for orchestras to present concerts online and if digital formats help orchestras to engage with new audiences.

The outcomes derived from the administered questionnaire show that with the support of digital formats, all participants have reached new audiences. In particular, younger audience demographics seem to have engaged with their digital concerts. The evidence of this study suggests that younger audiences seem to favour online events. These findings contribute to existing knowledge about audience demographics engaging with audio digital formats (see report mentioned in Section 6 published by the British Phonographic Industry, Deezer and Royal Philharmonic Orchestra in 2020).

Overall, orchestras seem to agree that digital formats have had a positive impact on audience engagement, in particular when orchestras were not allowed to perform live. Nonetheless, the results of this study support the idea that the experience for the audience of a digital concert is different from attending a live performance. Despite concerns about the social and communal aspect, the results demonstrate that digital formats have facilitated orchestras' engagement with audience segments who might otherwise abstain from attending live concerts (non-regular concertgoers who may feel intimidated by concert etiquettes, for example). These findings are significant in reaffirming the potential of digital formats in providing orchestras with audience retention and growth, which can be valuable for orchestras considering the expansion of their audience base. For example, if there has been an observed decline in the subscriber base.

Lastly, limitations need to be taken into account. First, the questionnaire considers the perspective of the orchestra but does not unveil the audience's experience when engaging with digital events. The majority of participants have shared feedback from their online concert audiences, which has been positive. However, it is essential to remain mindful of the potential for bias in these responses. It can be speculated, for instance, that marketing teams prefer not to share negative comments so as not to undermine the orchestra's reputation. Nevertheless, most of the audience feedback comes from social media platforms, which is accessible and thus possible to corroborate. Thereafter, an issue that was not addressed in this study was whether audiences who have engaged with digital concerts during the pandemic have thereafter started attending live performances. Most participants indicate that data were not available. Finally, it is unfortunate that the study did not reveal accurate data about audience demographics engaging with online events.

Further research is necessary to gain a more comprehensive understanding of whether digital formats can serve as a bridge between newer audience demographics and live concert performances. This inquiry holds substantial significance due to its alignment with Nicholls et al.'s (2018) contention that the demographic of classical music audiences is ageing. Consequently, orchestras are compelled to strategise for developing their future concert attendees.

While this study shows that digital formats have helped orchestras to engage with audiences during the COVID-19 pandemic, further research is needed to establish current rates of engagement. Future research investigating online engagement in orchestral contexts will determine whether digital formats constitute a viable opportunity for orchestras. Finally, another possible area of future research would be to assess audience demographics engaging with online events. This echoes the strategy of one participant, who suggests that more analysis of audiences engaging with their digital concerts is needed.

References

Airoldi, M. (2021). The techno-social reproduction of taste boundaries on digital platforms: The case of music on YouTube, *Poetics*, Vol. 23, No. 1, pp. 1–13.

Albright, C. (2016). 'Classical' music is dying . . . and that's the best thing for classical music, *CNN* (website), available online from https://edition.cnn.com/2016/05/29/opinions/classical-music-dying-and-being-reborn-opinion-albright/index.html [accessed 27 August 2023].

Arts Council England (2023). How orchestras can support our health and wellbeing, *Arts Council England* (website), available online from www.artscouncil.org.uk/blog/how-orchestras-can-support-our-health-and-wellbeing/ [accessed 20 August 2023].

Barna, E. (2020). The relentless rise of the poptimist omnivore: Taste, symbolic power, and the digitization of the music industries. In: Tófalvy, T. and Barna, E. (eds.). *Popular music, technology, and the changing media ecosystem: from cassettes to stream*, Cham, Springer International Publishing, pp. 79–93.

Barron, M. (2010). *Auditorium acoustics and architectural design*, 2nd edition, London, Spon Press.

Baym, N. K., and Boyd, D. (2012). Socially mediated publicness: An introduction, *Journal of Broadcasting & Electronic Media*, Vol. 56, No. 3, pp. 320–29.

Berger, B., Matt, C., Steininger, D. M., and Hess, T. (2015). It is not just about competition with "free": differences between content formats in consumer preferences and willingness to pay, *Journal of Management Information Systems*, Vol. 32, No. 3, pp. 105–128.

Borstlap, J. (2012). *The classical revolution: thoughts on new music in the 21st century*, Lanham, MD, Scarecrow Press.

Botstein, L. (2020). Post-pandemic anxieties: Contemplating the prospect that classical music culture might disappear. *The Musical quarterly*, Vol. 103, No. 1, pp. 1–8.

Bourdieu, P. (1984). *Distinction: A social critique of the judgement of taste*, Cambridge, MA, Harvard University Press.

Brace, I. (2013). *Questionnaire design: How to plan, structure and write survey material for effective market research*, 3rd edition, London, KoganPage.

Bradley, C. (2017). National Classical Music Audiences. An analysis of Audience Finder box office data for classical music events 2014–2016, *The Audience Agency (*website), available online from www.theaudienceagency.org/asset/1303#:~:text=The%20modelled%20age%20breakdown%20suggests,to%20be%20aged%20under%2031 [accessed 13 August 2023].

Braun, V., and Clarke, V. (2006). Using thematic analysis in psychology, *Qualitative Research in Psychology*, Vol. 3, No. 2, pp. 77–101.

British Phonographic Industry, Deezer and the Royal Philharmonic Orchestra (2020). *The classical revival in 2020: A report by BPI, Deezer and the Royal Philharmonic Orchestra*, available online from www.bpi.co.uk/media/2518/the-classical-revival-2020_final.pdf [accessed 20 August 2023].

Brown, A. S., and Novak, J. L. (2007). *Assessing the intrinsic impacts of a live performance*, available online from: www.culturehive.co.uk/wp-content/uploads/2013/04/ImpactStudyFinalVersion.pdf [accessed 29 August 2023].

Büscher, F. (2023). Zukunft Musikunterricht in Grundschulen. Kulturminister Piazolo signalisiert Zuversicht, BR Klassik (website), available online from www.br-klassik.de/aktuell/news-kritik/lehrermangel-und-musikunterricht-in-bayern-interview-kultusminister-michael-piazolo-100.html [accessed 13 August 2023].

Caldwell, M., and Woodside, A. (2003). The role of cultural capital in performing arts patronage, *International Journal of Arts Management*, Vol. 5, No. 3, pp. 34–49.

Chucherdwatanasak, N. (2020). Making Detroit sound great: The Detroit Symphony Orchestra and its post-strike transformations, *Artivate: A Journal of Entrepreneurship in the Arts*, Vol. 9, No. 1, pp. 43–62.

Colotti, C. (2021). A review of contemporary practices incorporating digital technologies with live classical music. In: Hepworth-Sawyer, R., Paterson, J., and Toulson, R. (eds.). *Innovation in music, future opportunities*, London, Focal Press, pp. 273–289.

Concerti (2016). concerti Klassikstudie 2016, *concerti Media GmbH* (website), available online from https://media.concerti.de/klassikstudie/ [accessed 13 August 2023].

Cooper, M. (2016). Metropolitan opera faces a slide in box-office revenues, *New York Times* (website), available online from www.nytimes.com/2016/05/07/arts/music/metropolitan-opera-faces-a-slide-in-box-office-revenues.html [accessed 20 August 2023].

Crawford, G., Gosling, V. K., Bagnall, G., and Light, B. A. (2014). An orchestral audience: Classical music and continued patterns of distinction, *Cultural Sociology*, Vol. 8, No. 4, pp. 483–500.

De Ridder, S., Vesnić-Alujević, L., and Romic, B. (2016). Challenges when researching digital audiences: mapping audience research of software designs, interfaces and platforms, *PARTICIP@TIONS*, Vol. 13, No. 1, pp. 374–391.

DeNora, T. (1997). *Music in everyday life*, Cambridge, Cambridge University Press.

Denscombe, M. (2017). *The good research guide: For small-scale social research projects*, 6th edition, Maidenhead, Open University Press.

Dillman, D. A., Smyth, J. D., and Christian, L. M. (2014). *Internet, phone, mail, and mixed-mode surveys: the tailored design method*, 4th edition, Hoboken, NJ, Wiley.

Dobson, C. M. (2010). New Audiences for classical music: The experiences of non-attenders at live orchestral concerts, *Journal of New Music Research*, Vol. 39, No. 2, pp. 111–124.

Dromey, C. (2018). Talking about classical music. Radio as public musicology. In: Dromey, C., and T. Haferkorn, J. (eds.). *The classical music industry*, Boca Raton, FL: Routledge.

Follansbee, J. (2006). *Hands-on guide to streaming media: an introduction to delivering on-demand media*, 2nd edition, Amsterdam, Elsevier/Focal Press.

Gans, H. J. (1998). What Can Journalists Actually Do for American Democracy? Harvard International Journal of Press/Politics, Vol. 3, No. 4, pp. 6–12. Available from https://doi.org/10.1177/1081180X98003004003 [Accessed 31 March 2022].

Gans, H. J. (1999). *Popular culture and high culture: an analysis and evaluation of taste*, Rev. & updated ed., New York, NY, Basic Books.

Gaylin, D. H. (2016). *A profile of the performing arts industry: Culture and commerce*, New York, NY, Business Expert Press.

Goulding, C. (2005). Grounded theory, ethnography and phenomenology: A comparative analysis of three qualitative strategies for marketing research, *European Journal of Marketing*, Vol. 39, Nos. 3/4, pp. 294–308.

Greckel, W. (1992). Visualization in the performance of classical music: A new challenge, *The Quarterly Journal of Music Teaching and Learning*, Vol. 3, No. 4, pp. 38–49.

Heilbrun, J., and Gray, C. M. (2001). *The economics of art and culture*, 2nd edition, New York, NY, Cambridge University Press.

Hutchby, I. (2001). Technologies, texts and affordances, *Sociology*, Vol. 35, No. 2, pp. 441–456.

Kavanagh, B. (2018). Reimagining classical music performing organisations for the digital age. In: Dromey, C., and T. Haferkorn, J. (eds.). *The classical music industry*, Boca Raton, FL: Routledge.

Kolb, B. M. (2005). *Marketing for cultural organisations*, 2nd edition, London, Thomson.

Kozinn, A. (1994). The Three Tenors, Guess Who, to Sing, *The New York Times* (website), available online from www.nytimes.com/1994/07/14/arts/the-three-tenors-guess-who-to-sing.html [accessed 15 August 2023].

Kramer, L. (2007). *Why classical music still matters*. Berkeley, CA, University of California Press.

Mackieson, P., Shlonsky, A., and Connolly, M. (2018). Increasing rigor and reducing bias in qualitative research: A document analysis of parliamentary debates using applied thematic analysis, *Qualitative social work*, Vol. 18, No. 6, pp. 965–980.

McCartney, K. F., and Jinnett, K. (2001). *A new framework for building participation in the arts*, Santa Monica, CA, Rand.

McEwen, L. (2015). Chinese orchestras seeks young fans for classical music, The Washington Post (website), available online from www.washingtonpost.com/goingoutguide/music/chinese-orchestra-aims-to-stay-relevant-but-not-stray-from-tradition/2015/12/09/adb71c3e-9a8c-11e5-94f0-9eeaff906ef3_story.html [accessed 13 August 2023].

Ministerio de Cultura y Deporte (2021). *Estadísticas de Artes escénicas y Música*, available online from www.culturaydeporte.gob.es/dam/jcr:1feaebe0-1303-4c9f-af12-f71e383cf2c3/datos-estadisticos-artes-escenicas.pdf [accessed 13 August 2023].

Naveed, K., Watanabe, C., and Neittaanmäki, P. (2017). Co-evolution between streaming and live music leads a way to the sustainable growth of music industry – Lessons from the US experiences, *Technology in Society*, Vol. 50, pp. 1–19.

Nicholls, C. D., Hall, C., and Forgasz, R. (2018). Charting the past to understand the cultural inheritance of concert hall listening and audience development practices, *Paedagogica Historica*, Vol. 54, No. 4, pp. 502–516.

O'Sullivan, T. (2009). All together now: A symphony orchestra audience as a consuming community, *Consumption Markets & Culture*, Vol. 12, No. 3, pp. 209–223.

Peterson, R. A., and Simkus, A. (1992). How musical tastes mark occupational status group. In: Lamont, M., and Fournier, M. (eds.). *Cultivating Differences*, Chicago, IL, University of Chicago Press, pp. 152–186.

Peterson, R. A., and Kern, R. M. (1996). Changing highbrow taste: From snob to omnivore, *American Sociological Review*, Vol. 61, No, 5, pp. 900–907.

Peukert, C. (2019). The next wave of digital technological change and the cultural industries, *Journal of Cultural Economics*, Vol. 43, No. 2, pp. 189–210.

Pitts, S. E., Dobson, M. C, Gee, K., and Spencer, C. P. (2013). Views of an audience: Understanding the orchestral concert experience from player and listener perspectives, *Participations: Journal of Audience and Reception Studies*, Vol. 10, No. 2, pp. 65–95.

Presidenza del Consiglio dei Ministri [2020] Modulario P. C. M. 198 MOD. 3 ART. 1.

Prime Minister's statement on coronavirus (COVID-19) (2020). Prime Minister Boris Johnson addressed the nation on coronavirus, [transcript of the speech]. *BBC* (website), available online from www.gov.uk/government/speeches/pm-address-to-the-nation-on-coronavirus-23-march-2020 [accessed 13 August 2023].

Psathas, G. (1973). Introduction. In: Psathas, G. (ed.). *Phenomenological sociology: Issues and applications*, New York, NY, Wiley, pp. 1–21.

Radio Joint Audience Research (2022). Quarterly Summary of Radio Listening, *Radio Joint Audience Research* (website), available online from www.rajar.co.uk/docs/2022_12/2022_Q4_Quarterly_Summary_Figures_YOY.pdf [accessed 20 August 2023].

Ramnarine, T. K. (2011). The orchestration of civil society: Community and conscience in symphony orchestras, *Ethnomusicology Forum*, Vol. 20, No. 3, pp. 327–351.

Rayburn, D. (2005). *The business of streaming and digital media*, Amsterdam; Burlington, MA, Focal Press.

Roose, H. (2008). Many-voiced or unisono?: An inquiry into motives for attendance and aesthetic dispositions of the audience attending classical concerts, *Acta Sociologica*, Vol. 51, No. 3, pp. 237–253.

Royal Philharmonic Orchestra (2023). *A time to look forward: Trends of engagement with orchestral music*, available online from www.rpo.co.uk/images//articles/2023_Insights_Report/RPO_Insights_Report_2022_release.pdf [accessed 21 August 2023].

Rußell, R., Berger, B., Stich, L., Hess, T., and Spann, M. (2020). Monetizing online content: Digital paywall design and configuration, *Business & Information Systems Engineering*, Vol. 62, pp. 253–260.

Schutt, R. K. (2006). *Investigating the social world: The process and practice of research*, 5th edition, Thousand Oaks, CA, Sage.

Stauber, R. (2022). Die Junge Musiker und das alte Publikum, Berliner Morgenpost (website), available online from www.morgenpost.de/kultur/article236464409/Die-jungen-Musiker-und-das-alte-Publikum. html [accessed 13 August 2023].

Szabó, R. Z., and Szedmák, B. (2020). The value innovation of symphony orchestras and the triggering effect of coronavirus. *Theory Methodology Practice (TMP)*, Vol. 16, No. 2, pp. 89–95.

Taruskin, R. (1997). *Defining Russia musically: Historical and hermeneutical essays*, Princeton, NJ, Princeton University Press.

Tepper, S. J., and Hargittai, E. (2009). Pathways to music exploration in a digital age, *Poetics*, Vol. 37, No. 3, pp. 227–249.

Thompson, S. (2007). Determinants of listeners' enjoyment of a performance. *Psychology of Music*, Vol. 35, No. 1, pp. 20–36.

Trainer, A. (2015). Live from the ether: YouTube and live music video culture. In: Cresswell, J. A., Bennett, R. J., and Jones, A. (eds.). *The digital evolution of live music*. Oxford, Elsevier Science & Technology, pp. 71–84.

Tunstall, T. (2013). Another Perspective: El Sistema–A Perspective for North American Music Educators, *Music educators journal*, Vol. 100, No. 1, pp. 69–71.

Turrini, A., O'hare, M., and Borgonovi, F. (2008). The border conflict between the present and the past: Programming classical music and opera, *The Journal of Arts Management, Law, and Society*, Vol. 38, No. 1, pp. 71–88.

Varney, D., and Fensham, R. (2000). More-and-less-than: Liveness, video recording, and the future of performance. *New Theatre Quarterly*, Vol. 16, No. 1, pp. 88–96.

Vladica, F. and Davis, C.H. (2013). Value propositions of opera and theater live in cinema. World Media Economics & Management Conference. Thessaloniki, Greece. 23-27 May 2013. Available from https:// people.ryerson.ca/c5davis/publications/Vladica-Davis%20-%20value%20propositions%20of%20l ive%20cinema%20-%20l%20May%202013.pdf [Accessed 10 June 2023].

Vroon, D. (2014). *Classical music in a changing culture: essays from the American Record Guide*, Plymouth, England, Rowman & Littlefield.

Walmsley, B. (2016). From arts marketing to audience enrichment: how digital engagement can deepen and democratize artistic exchange with audiences, *Poetics*, Vol. 58, pp. 66–78.

Watts, S. (2014). User skills for qualitative analysis: Perspective, interpretation and the delivery of impact, *Qualitative Research in Psychology*, Vol. 11, No. 1, pp. 1–14.

Webster, J. (2020). Taste in the platform age: music streaming services and new forms of class distinction, *Information, Communication & Society*, Vol. 23, No. 13, pp. 1909–1924.

8

WHEN CHORD CHARTS FAIL

Pitfalls of Radical Reharmonisation of Jazz Standards

Agata Kubiak-Kenworthy

1. Introduction

The motivation behind this research project arose from past practical problems in communicating complex modern jazz harmony that the author often encountered in their career as a performing jazz musician and a band leader. The form of a chord chart is still the most common way of communicating original modern jazz music as well as original arrangements of jazz standards. The prevalence of the use of applications like iReal Pro only cemented the departure from the Western classical music notation in the context of jazz performance.

When producing a chord chart for a jazz composition, a band leader has a difficult job of communicating their compositional language as well as accommodating all band members' preferred ways of seeing the harmony laid out in a chord chart. Some pianists really value the use of slash chords, and some prefer modes. Some instrumentalists would like to see all the chord extensions 'spelled out' and some would rather see the most essential ones only. Some drummers prefer not to see the chords and the melody in order to focus on the structure, feel and potential hits, and some would feel very patronised if you presented them with a 'bare' drum chart. The author's intention is to take a first step into discovering and formalising some of these instrument-specific preferences while questioning whether the form of a chord chart is the best way of communicating modern jazz harmony that drifts away from its functional roots.

2. From Tune-Dex Cards to iREAL Pro

Jazz is an aural tradition at its core. The drive of jazz studies in becoming an academic discipline resulted in the gradual departure from its aural roots. The journey of formalisation of the jazz language began in the 1950s with George Russell's 'Lydian Chromatic Concept of Tonal Organization' (Russell, 1953). Later efforts in compiling the practical knowledge were presented by Jerry Coker in his *Improvising Jazz* (Coker, 1964) and *Patterns for Jazz* (Coker, 1970). Following these, the later works of Mark Levine (Levine, 1989), (Levine, 1995) and Berklee College of Music's 'chord scale theory' (Graf, Nettles, 1997) have a direct impact on how we teach, study and play jazz today.

Lead sheets and later chord charts, commonly used in performances by jazz practitioners today, originated in the 1940s as George Goodwin's Tune-Dex cards. This was a subscription

DOI: 10.4324/9781003396550-8

service for radio stations, music professionals and musicians which allowed them to keep track of popular songs of the time (Kernfeld, 2003). Lead sheets provide the melody, lyrics and the chord progression for each song, whereas chord charts only provide the chord progression in a given key. Compilations of jazz lead sheets, referred to as Fake Books, have been around since the 1920s (Kernfeld, 2006). The transformation into what is known as 'The Real Book' took place in 1975. The creation of this seminal collection of lead sheets is attributed to several anonymous Berklee College of Music students. The young musicians were tired of the illegible and somewhat irrelevant fake books, available to music students at the time, set about transcribing popular jazz tunes pertinent to the time into handwritten charts. Riffing on the title of a popular Cambridge street paper called the Real Paper and the concept of taking "fake" books to the next level, the Real Book was born (*Official Real Book, 2023*). Regardless of the not-for-profit intentions behind the creation of the Real Book, the publication remained illegal till 2004, when the publisher Hal Leonard took it over while securing all the necessary copyrights. The Real Book has been a worldwide success and a practice and gig companion to many jazz students and professionals for almost 50 years.

2010 saw the arrival of an app called iReal Pro. Designed by a New York-based jazz musician – Massimo Biolcati – it became appreciated very quickly. The app, listed by *Time* in '50 Best Inventions of 2010', is currently used by 2 million+ musicians worldwide (irealpro.com, 2023)

The current wide use of applications such as a chord charts library 'iReal Pro' drastically reduced the use of lead sheets.

3. The 'Smile' Experiment

This chapter presents the idea and motivation behind The 'Smile' Experiment, its set-up, participants and the discussion of results. Full arrangement as well as audio examples of a piano and guitar follow-up experiment described later in this chapter can be listened to here: https://on.soundcloud.com/uJLBq

3.1. *Motivation and Initial Set-up*

As mentioned in the Introduction to this paper, the motivation behind creating a first iteration of this study arose from some professional frustrations and struggles in communicating compositional material in a modern jazz setting. The socio-culturally expected use of chord charts as a medium of notation seemed to limit the harmonic content of compositional ideas that were being communicated and often created misunderstandings in rehearsal and live performance settings. Every time a new or a deputy band member would join the group, a lot of time had to be spent on clarifying the intention of the composer/bandleader. The author decided to follow the hint from their own professional practice and design a study that could potentially showcase the level of potential musical misunderstandings and pitfalls of using the established jazz notation in communicating bordering non-functional, original, modern jazz harmony.

The initial setup of the experiment involved creating and recording a radical reharmonisation of a popular jazz standard 'Smile'. The choice of this particular song was justified practically and artistically. The practical motivation was due to popularity of the jazz standard and a belief that it was highly unlikely for a professional jazz musician to be unfamiliar with the composition. It was relevant that the original chords featured in the song are rather simple which would potentially allow for more adventurous 'tempering' with the harmony while creating the arrangement. The artistic reasons behind the song choice were connected to the lyrical material. The author felt that an arrangement which emphasised the darker side of the song would not be out of place in the esthetical sense. This interpretation stylistically suited compositional material formerly engaged with by the author, which made the arrangement an artistic statement as well as a research experiment. The professional recording of the arrangement (accessible here: https://on.soundcloud.com/uJLBq) is to be released in early 2024 on the author's upcoming second album 'The Other'.

The arrangement was written by the author in 2020 and the first version of the recording took place at the author's home studio during one of the lockdowns. The instrumentation chosen for the arrangement consisted of: 2 violins and 2 violas, a solo violin and voice. The arrangement was then re-recorded in a professional studio in 2022 with the same instrumentation and following the same structure and chords. The latter version was the one used in this research.

There were several reasons behind the unusual choice of instrumentation. As the first recording took place in lockdown, access to musical collaborators was highly limited. The author performs on all the instruments used up to a professional standard, so it was an easy solution to create a self-sufficient arrangement at the time. String instruments were also heavily featured on the author's debut album 'Polarity', which made this set-up a familiar creative environment. There was another, more cerebral reason behind using this group of instruments. The author did not want to suggest any familiar chord voicings to potential future research project participants. The worry that use of a piano, guitar or bass could (consciously or not) influence participants' responses, merited and cemented the unusual instrumentation choice.

While the harmony of 'Smile' was radically changed, the melody remained almost entirely resembling the original. This conservative treatment of melody and lyrics was premeditated and played the role of a balancing musical component, which complemented and softened the oppositional harmonic material.

The visual presentation of the score (Figure 8.1) was also an important decision for the author, prior to conducting the study with chosen participants. As the author believed that it is unlikely that the participants might be unfamiliar with the original melody of the standard, a choice was made not to transcribe the melody onto the score. This was also to channel the 'iReal Pro experience' of only looking at a chord chart, which became a standard practice among jazz practitioners in the last 13 years. Another important aspect of the presentation of the score was the lack of key signatures. The decision to use accidentals instead of a tonal centre or a key was made to create a possibility of a discussion around the tonality of the arrangement. However to counter any possible misunderstanding of how the melody and chords interacted, a full audio recording of the arrangement was also provided to each participant.

The radical reharmonisation, measuring 34 bars, was created by displacing some of the original chord tones, while keeping some. New chords were then voiced in a specific way to disguise original chord tones. The bordering non-functional sound of the harmony is surprising once one realises how much shared material there is between the original and reharmonised chords.

Figure 8.2 presents the connection with the original chords. Notes highlighted in red represent the notes in common and the Roman numerals below state which chord tone was kept from the original harmony (I – root, III – third, V – fifth and VII – seventh). There were only two instances where just one chord tone was kept – bar 13 and bar 29. The majority of chords have two chord tones in common, and bars 3, 4, 8, 9, 10, 19, 20, 24, 25, and 26 all have as many as three chord tones in common. It is also worth noting that as many as 26 bars contain a root of the original chord. On the other hand, as many as 20 bars contain the fifth of the original chord as a kept note, which is another trick used to lead the listener's ear astray. The fifth of the chord is often the first note to be discarded if one tries to strip the chord of the 'unnecessary' material. It is not uncommon for pianists to strip chords down to the third and seventh while engaged in comping.

Smile

arr Agata Kubiak-Kenworthy

Charles Chaplin

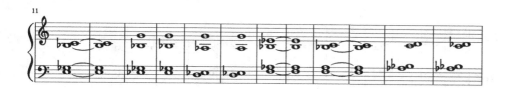

FIGURE 8.1   Full arrangement as used in the Smile Experiment.

3.2. Participants and Feedback

Participants were selected from the author's professional network. Some participants responded after being messaged personally and some got in touch after noticing a 'Call for participation' published on social media by the author. The initial study consisted of 16 participants including: 5 pianists, 2 drummers, 3 bass players, 2 guitarists and 4 lead instrument players (vocalist, trumpeter and 2 saxophonists). All participants are professional jazz musicians, active within their field.

The following simple instruction was given to each willing participant:

> If you could please write a jazz chord symbol in each text box. One that best describes the dots in your opinion. If you would like to listen to the arrangement, here it is: (link to soundcloud).

They were all provided with the chord chart written out in notation (*Figure 8.1*) and a link to the audio recording of the full arrangement. Participants were instructed to contact the researcher if

Smile

arr Agata Kubiak-Kenworthy

Charles Chaplin

FIGURE 8.2 Full arrangement highlighting the chord tones in common with the original harmony.

the task was not clear, and they were also asked if they would like to remain anonymous. None of the participants expressed the need for anonymisation but they will be referred to by their initials and instrument as follows:

- B.C. (piano)
- R.B. (piano)
- D.V. (piano)
- L.M. (piano)
- A.B. (piano)
- L.B. (drums)
- D.B. (drums)
- J.M. (bass)
- C.P. (bass)
- S.P. (bass)
- P.B. (guitar)
- T.K. (guitar)
- A.C. (voice)
- D.M. (trumpet)
- L.P. (saxophone)
- B.D. (saxophone)

No feedback about the process was specifically requested from participants at this stage of the study. Regardless, some musicians provided feedback, and some responded with jokes that betrayed how uncomfortable they found the task. D.V. (piano) joked: 'I thoroughly hated doing it and what I've produced is the result of frustration and despair', while P.B. (guitar)'s humour went even further with the statement: 'You are clearly mentally ill!' Pianists were particularly vocal about the difficulty of the task. R.B. (piano) stated: 'I wouldn't be comfortable putting ANY chords on small clusters', while B.C. (piano) wrote: 'What you sing over them cancels them out anyway'. D.V. (piano) elaborated a bit more: 'I genuinely had no clue for some and quite possibly all . . . Seemingly impossibly trying to balance readability vs some sort of actual truth'. The author also received a comment from C.P. (bass): 'I didn't include all the 9ths I might have done . . . typical bass player approach!' and another comment from D.M. (trumpet): 'I could have maybe thought more about how the chord sounds other than its functioning when naming . . . '. These last two betray some more stereotypical approaches connected with the instrument in question. D.M. (trumpet) also admitted that the lack of focus on the sound was because he could only listen to the arrangement once, before completing the task. Imagining the sound of the harmony is difficult for lead instrument players.

3.3. Results and Discussion

Just after the first few responses the author could see that different approaches were being taken by different musicians. Some participants preferred to see a lot of detail (for example one pianist's 'Abmaj7, ommit3/D' in response to bar 15), while others preferred a more sparse approach (for example a different pianist's 'Gsus' in response to bar 8). D.V. (piano) was the only participant who decided to use names of modes to describe the chords, which made for an interesting individual approach, neither forbidden nor encouraged by the author in the description of the task. The most disagreement behind what the root of the chord was could be seen in responses by the 5 pianists, while the other musicians' responses were more consistent overall.

The author did not specifically request participants to listen to the arrangement, but the hope was that when needed, it could be referred to for guidance. It became apparent that this was not clear when one of the participants, L.M. (piano), contacted the author saying: 'I wasn't sure what key I was in but I would guess Eb major', when the key of the arrangement was in fact Bb. This would have been very clear to anyone who listened to the arrangement as the melody was very much unchanged. Another way of being sure that not all participants listened to the arrangement was the count of the Soundcloud link provided. The file was made private, so only a person with the link could play it. After receiving all the responses the author noted 'the listed to' count as 12 for 16 participants. It is hard to argue that hearing the arrangement would not potentially be helpful so the question of whether it was ignored on purpose or accidentally by some participants remains unanswered. This is one of the reasons why the author decided to implement a post-participation questionnaire for the next iteration of this research project.

The author was interested in finding which chords in the arrangement were the most agreed and disagreed upon. After analysing all 16 responses, it became clear that the most agreed-upon chord was the one featured in bar 3 (from bottom to top: F, A, Bb, C). As many as 15 participants identified this chord as F(add4). The exception came from C.P. (bass) who identified this chord as Bbmaj/F.

There were two chords worth mentioning on the opposite side of the spectrum. First, the most disagreed chord in the arrangement was the one featured in bar 16 (from bottom to top: D, Eb, A, G). In this case, as few as 2 participants agreed on the chord name and type. It makes it particularly impressive when considering these 4 notes (2 of which are 3rd and 7th of the original harmony) could be described in 15 different ways by 16 musicians. It was clear what sound the author was looking for a description of. The context of the chord was also known and the resemblance to the original harmony was as close as it could, without also adding a root note, but somehow the voicing of the chord and the choice of two remaining notes made the task very difficult. Another chord which caused a lot of disagreement was the one featured in bar 8 (from bottom to top: F, G, C, D). This clustered voicing included as many as 3 common notes with the original chord and it produced 13 different versions of assigned chord symbols from participants.

The tremendous variety of some of these responses inspired the author to create another stage of this initial study, later referred to as 'Guitar/Piano follow-up experiment'. The author gathered four of the most distant chord symbol responses (describing the same four notes) from bars 8, 11 and 16. Those were then sent to one pianist (R.B.) and one guitarist (T.K.) who were asked to produce a simple recording of each of these chords. The information musicians were given on this occasion was very sparse. They were not made aware that each four chords were to describe one (the same!) sound. They were simply just requested to play what they saw on the page and create one sound file per example. The author encourages the reader to listen to those examples (accessible here: https://on.soundcloud.com/uJLBq) as the results are quite surprising. Needless to say, there was little resemblance between the sounds of four chords in each of the examples. There were also very noticeable differences to how the chords were executed by the pianist and the guitarist, but this was to be expected due to limitations of voicings on respective instruments.

The author was also interested in exploring the potential differences in slash chord (for example Dbmaj7/Ab) and sus chord (for example C7sus4) use within different instrument groups. The appearance of both slash chords and sus chords in each instrumental group was counted and presented in a form of a pie chart (Figure 8.3 and Figure 8.4).

Preference for slash chord use was clear in bass players' responses. This was to be expected, once taken into consideration the specific role of the instrument. The author, however, was surprised by a rather minor difference between piano and guitar players. This was probably due to a very small sample size. The author's prediction would have been that a slash chord would be a rather popular response among pianists as it is often treated as a very 'piano-friendly' way of 'spelling out' the harmony. The author decided to ask a follow-up question regarding slash chord use to one of the guitarists involved with the study. The participant responded stating his preference for as few slash chords as possible in the chord chart and proceeded to explain that at its extreme it is similar to handing a guitar tablature to a piano player. 'I'm not interested in seeing what the pianist's hands are doing' – he concluded.

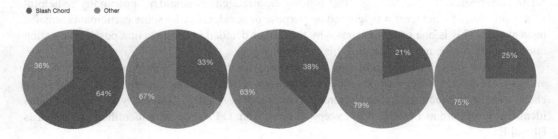

SLASH CHORD USE (from left to right): Bass, Piano, Lead Instruments, Drums, Guitar

FIGURE 8.3 Prevalence of Slash Chord use in 5 instrumental groups.

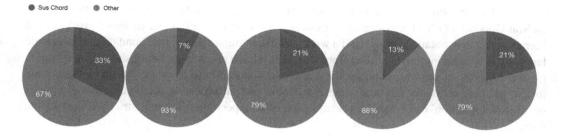

SUS CHORD USE (from left to right): Bass, Piano, Lead Instruments, Drums, Guitar

FIGURE 8.4   Prevalence of Sus Chord use in 5 instrumental groups.

The sus chord use followed a similar trend as slash chords and the pianists' lack of preference here was even more prominent. As before, the bass players stated the most preference for sus chord used and the drummers the least (just as with the slash chord used above). The clear distinction between pianists and lead instrumentalists in both cases of slash and sus chords used is of interest here and will be followed up in the future iteration of this study. It is fair to notice this aspect of the study could especially benefit from a larger sample, when some more significant patterns could potentially be observed. A post-participation questionnaire could also look deeper into personal preferences and instrument-specific scoring habits.

3.4.   Next steps

As mentioned in the discussion, the author would like to continue work on this topic. The next iteration of the study is to include at least 10 participants per instrument group as well as pre-participation interview and post-participation questionnaire. The next run of the study will also include a practice element, where participants will be requested to record a short (1-2 choruses) solo while using the chart they wrote. Potential part 3 of the study could include 'a chart swap' where musicians record a solo while reading from someone else's chart. The latter has not been decided on yet as it might create too many variables to analyse.

4. Conclusion

This study looked into how different groups of professional jazz musicians approach generating a chord chart to a radically reharmonised jazz standard. 16 participants (5 pianists, 3 bass players, 2 drummers, 5 lead instrumentalists and 2 guitarists) were selected to participate. Musicians were requested to produce a chard chart of the most appropriate chord symbols. The materials provided were a staff-notated arrangement (Figure 8.1) and an audio recording. The results were suggestive of major differences in understanding the harmony. The culmination of which was bar 16, where 15 different versions of a chord symbol were used to describe the same 4 notes. If 16 professional jazz musicians are capable of producing 15 different versions of a chord symbol in order to describe a sound of four notes, then surely the format of a chord chart is not the best medium in communicating harmonic compositional language . . . Some of those chords will of course be a different way of stating the same (or similar enough) sound, but not all. The Guitar-Piano follow-up experiment (https://on.soundcloud.com/uJLBq) shows what array of different sound worlds a composer could end up with, just as a result of interpretation of the same 4 notes. One might wonder whether we should worry about such issues in an improvisation-heavy medium such as jazz. Philosopher Nelson Goodman (1969) states that: 'a musical score is in a notation and defines a work, . . . a literary script is in notation and is itself a work'. The communication of the foundation of harmonic language in a jazz composition is therefore crucial as it has a potential to establish a strong common ground towards an improvisation on the piece within an ensemble setting. Using staff notation instead of chord charts for this task seems reasonable, but unfortunately socio-cultural associations of 'classical notation' can create intimidation and false expectations around 'strictly following the score'.

To conclude with words by Tim Ingold:

But the composer does not write a musical work. He writes a score, which in turn specifies a class of performances compliant with it. The musical work is that class of performances.

(Ingold, 2016, p. 11)

References

Coker, J. (1964, rev. 1986). *Improvising Jazz*, Simon & Schuster.
Coker, J. (1970). *Patterns for Jazz*, Alfred Music Publishing.
Goodman, N. (1969). *Languages of Art: An Approach to a Theory of Symbols*, Oxford University Press.
Graf, R., Nettles, B. (1997). *The Chord Scale Theory and Jazz Harmony*, Advance Music.
Ingold, T. (2016). *Lines*, Routledge.
IReal Pro, http://irealpro.com (accessed 28th of August 2023).
Kernfeld, B. (2003). *Pop Song Piracy, Fake Books, and a Pre-history of Sampling*, University of Chicago Press.
Kernfeld, B. (2006). *The Story of Fake Books – Bootlegging songs to musicians*, Rowan and Littlefield.
Levine, M. (1989). *The Jazz Piano Book*, Sher Music Co.
Levine, M. (1995). *The Jazz Theory Book*, Sher Music Co.
Official Real Book, http://officialrealbook.com/history/ (accessed 28th August 2023).
Russell, G. (1953). *Lydian Chromatic Concept of Tonal Organization*, Concept Publishing.

9

DIGITAL AESTHETICS AND TRANSCENDING LO-FI IN ALEX G'S *GOD SAVE THE ANIMALS*

Jamie Birkett

1. Introduction

The fetishization of analogue technology is seemingly hard-set into the recording industry at this point; the sound of the recorded guitar music of the 1960s and 70s is often presented as an ideal of authentic aesthetic fidelity.

Historically, indie-rock as a genre had both embraced and transgressed these values by creatively utilizing consumer-level recording technologies and home-recording techniques, resulting in the codification of 'lo-fi' in the 1990s and a subsequent aestheticization of the sound of consumer tape distortion and conventionally 'bad' sounding home recordings; as exemplified by the likes of Guided by Voices, Pavement and Duster. And yet the consumer recording technologies of today offer a transparent fidelity that could not be achieved even in the best studios of the 'golden age'. Contemporary indie-rock records may well make extensive use of modern technologies, all while trying to recreate the aesthetic appearance of recordings made in the analogue era.

In Alex G's *God Save The Animals* (2022) we hear a sincere attempt to embrace the affordances and immediacy of contemporary recording technologies over the fetishization of pre-internet indie-rock aesthetics which still haunt the contemporary sound-world. The record sounds like a product of the post-internet world where any music is easily accessible, and musical identity is more fluid. We are met with the odd juxtaposition of rootsy folk-rock alongside hyper-pop, slow-core and SoundCloud rap, emo-revival, and nu-metal. Bad taste is embraced, subverting expectations of 'good' and 'tasteful' record production.

This paper is primarily attempting to articulate the significance of creative choices made in what Moylan (2020) calls the domain of 'Recording', i.e. "the sound qualities and interrelations of sound qualities that are brought about by the recording process". Considering contemporary trends of analogue fetishism in contemporary guitar music, *God Save The Animals* is presented as a case study of a work which embraces 'the digital' as an aesthetic value that transgresses the stylistic conventions of indie-rock. The study takes methodologies outlined in Moylan's *Recording Analysis* (2020) as applied to track-by-track textual analysis. The methodology is used to parse the significance of notable production choices and sonic markers, all of which ultimately combine to create an aesthetically consistent sound-world from a diverse and seemingly divergent range of styles.

DOI: 10.4324/9781003396550-9

2. Analogue Fetishism and Retromania

The notion of 'analogue fetishism' draws upon Karl Marx's theory of commodity fetishism which he sees as the "mistaken view that the value of a commodity is intrinsic" (Buchanan, 2010) and that certain commodities can be given a culturally symbolic value which appears to supersede the objective qualities of the object.

In our case, we may see this as analogous to how the sonic qualities imparted by 'vintage' analogue recording technologies are still considered desirable characteristics with an aesthetic value which belies the measurable fidelity of a digital recording. This tendency may be recognized in the digital tools which are modelled to replicate the technical 'imperfections' of older physical technologies. Moreover, the symbolic value of analogue recording technology is not limited to audible qualities; objects can also be valued for other characteristics such as physical tactility, historical associations, mythology, rarity, etc.

We can better understand the notion of analogue fetishism by looking to Paul Théberge's seminal study of music technology and consumption, *Any Sound you Can Imagine* (1994). He draws upon the theorist William Leiss' idea of commodities as "material-symbolic" entities with "imputed characteristics":

> Commodities have reached a level of complexity during the twentieth century that requires consumers no longer to judge them purely on their objective properties but also on the particular "characteristics" they embody. Advertising, socialization patterns, and interpersonal relations all play a role in determining the imputed characteristics of commodities.
>
> *(Théberge, 1994, p.139–140)*

Tangled up in a post-modern tendency towards what critic Simon Reynolds (2012) has called Retromania, or "pop culture's addiction to its own past", we can see clear examples of this symbolic value given to vintage recording technologies. It is present in the mythologies created around the recording process in cases such as the retro-revivalist projects of Jack White's Third Man Records and in Dave Grohl's *Sound City* documentary, in which vintage analogue recording technologies are often presented as an ideal of authentic aesthetic fidelity and 'realness'. In these instances, it is the physical objects that provide a material conduit to pass down extra-musical stories, as if the object is imbued with the spirit of the music itself, something that is seemingly less plausible with the inherently duplicable non-materiality of digital plug-ins.

These kinds of stories are still being told through the recording process and subsequent media cycle today. In 2022, the same year that *God Save The Animals* was released, we saw notable records within the same indie/folk-rock idiom such as Weyes Blood *And in The Darkness Heart Aglow* (2022) playing with nostalgia for 1960s recording aesthetics, produced by Jonathan Rado in the legendary Studio 3 at EastWest in Hollywood. Engineer and producer Sam Evian famously recorded highlights from Big Thief's *Dragon New Warm Mountain I Believe in You* (2022) on his front porch with a 4-track tape machine 'jerry-rigged' to his truck's cigarette lighter during a power outage (4ad, 2021).

Within the conjuncture of analogue fetishism and a trend towards sonic nostalgia, the logical progression of these aesthetic values is what music critics have jokingly referred to as 'faux-fi'. In reference to the latest album *V* by Unknown Mortal Orchestra. Pitchfork reviewer Sadie Sartini Garner (2023) describes how

> sweet vocals sung through grimy mic filters, songs squeezed with ear-popping compression—became one of the defining aesthetic markers of the indie rock of the last decade; the clipped

and redlining drum sound of a UMO type beat will telegraph 2010s bedroom rock to future generations in the same way gated reverb connotes the megahits of the 1980s.

In Retromania (2011), Reynolds describes how the recordings of Ariel Pink and others of the post-internet indie-rock era sought to recreate the characteristics of older styles in fine detail. Yet it is interesting that this pastiche of retro recording aesthetics became so commonplace that it developed beyond a simple re-creation of pre-existing styles into an identifiable style of production in and of itself. This has the peculiar effect of rooting the music within a contemporary sound-world that will itself be recognizable to future generations as distinctly of its time.

3. Background

Alex Giannascoli, better known as Alex G, is an American singer, songwriter and producer based out of Philadelphia, PA. He "built his reputation with an intimate lo-fi pop that combined strong melodic sensibilities and intimate vocals with (. . .) diversions into recording experiments and noise" (Donelson, no date).

Giannascoli's recording career started out as a teenager uploading self-released home recordings to the online music platform Bandcamp in the early 2010s. A reserved and private artist, there was a perceived mystique to his online presence at this time. His particular combination of pop songwriting, lo-fi recording and a playful experimentalism quickly garnered him a devoted and often obsessive fanbase. This fanbase is known for their attempts to uncover rare recordings during this prolific early period; the website unreleasedalexg.com (2021) is devoted to fan-made compilations of unreleased Alex G material that had originally been distributed unceremoniously via CDR and email. There is also a 20,000+ strong subreddit community devoted to discussion and speculation on the minutiae of Giannascoli's music; to the point that fans have expounded theories on how each song is written from the point of view of a different animal, or how every song is about a dog.

After releasing nine full-length records over the last twelve years and signing to Domino Records for the 2015 release of *Beach Music*, he continued to self-produce and record at home up until and including 2019's *House Of Sugar*. Although initially hesitant, he had been encouraged by the label to collaborate with *Unknown Mortal Orchestra* bassist and producer Jacob Portrait who had been brought in to mix the first three Domino releases. For *House Of Sugar* (2019) Giannascoli made use of a single U87 clone and an updated copy of GarageBand. According to Portrait, this was a marked upgrade in comparison to the even more rudimentary recording setup that Giannascoli had used for every record previously; "He would just record everything into GarageBand through a $90 USB microphone that he connected with a cable directly into his laptop" (Ducker, 2022).

All this is to say that the quality of the recording and lo-fi aesthetic had always been an integral part of Alex G's music. There have been multiple online media outlets that published pieces describing him as "the internet's secret greatest songwriter" (McDermott, 2014), and it seems the juxtaposition of astutely skilled pop songcraft with an irreverence for professionalism all amounted to the appeal and mythology of the artist; the prolifically inspired prodigy making music in his bedroom, underappreciated and unaware of his own genius.

Giannascoli's approach to recorded music is one example of the immediacy that contemporary home production affords; the distinction between songwriting and recording processes are collapsed, becoming one and the same process. The capture and creation of sound texture becomes embedded as a fundamental material of the composition unfolding in real time. Any separation between the writing and recording process is no longer necessary.

We can place the recording techniques of Alex G within the lineage of a DIY recording aesthetic which can be traced back to the low-budget independently released punk and post-punk records of the late 1970s. Indie-rock as a genre had both embraced and transgressed these values by creatively utilizing consumer-level recording technologies and home-recording techniques, resulting in the codification of 'lo-fi' in the 1990s and a subsequent aestheticization of the sound of consumer tape distortions and conventionally 'bad' sounding home recordings.

But notably, *God Save The Animals* is his first record to step outside of the bedroom environment, and much was made in the album press cycle of the fact that this was his first 'studio album':

> As with his previous records, Giannascoli wrote and demoed these songs by himself, at home; but, for the sake of both new tones and "a routine that was outside of my apartment," he asked some half-dozen engineers to help him produce the "best" recording quality, whatever that meant. The result is an album more dynamic than ever in its sonic palette.
>
> *(Bandcamp, 2022)*

It is the careful crafting of this 'dynamic sonic palette' which is an essential feature of the production style of *God Save The Animals*, and the source of much debate and discussion for many media outlets and online commentors. The unexpected and seemingly incongruent assemblage of disparate styles and sonic markers is reminiscent of what new media artist and theorist Marisa Olson coined as 'post-internet' art as early as 2006:

> What I make is less art 'on' the Internet than it is art 'after' the Internet. It's the yield of my compulsive surfing and downloading. I create performances, songs, photos, texts, or installations directly derived from materials on the Internet or my activity there.
>
> *(Olson in Connor, 2014)*

God Save The Animals often sounds like a product of a post-internet artist attempting to make sense of the information overload experienced by the contemporary listener when faced with the combination of near infinite access to the music of yesteryear alongside a proliferating stream of online microgenres.

The following track analysis attempts to unpack the significance of notable production choices which ultimately combine to create an aesthetically consistent sound-world from a diverse and seemingly divergent range of styles.

4. Track Analysis

4.1. 'After All'

The album opens with the track 'After All', which acts almost like an overture to introduce many of the recurrent themes of the record. We first hear a fairly conventional folk-rock arrangement, comprising of what could be identified as the 'core' instrumentation making up most of the record: acoustic rhythm guitar, piano, bass guitar, modulated electric 'lead' guitar, and 'live' drum kit.

The track takes an unconventional turn with the introduction of the vocals which are heavily processed; pitch-shifted up and treated with a large hall reverb. This production choice I believe is of particular significance to the lyrical themes of the record. The multiple layers of pitched-up and 'washed-out' vocals recall the sound of a children's liturgical choir – echoing the prominent themes of God and religion that permeate the record:

"God" figures in the ninth album from Philadelphia, PA based Alex Giannascoli's LP's title, its first song, and multiple of its thirteen tracks thereafter, not as a concrete religious entity but as a sign for a generalized sense of faith (in something, anything) that fortifies Giannascoli, or the characters he voices, amid the songs' often fraught situations.

(Bandcamp, 2022)

The infantilizing effect of the pitch-shifting combined with a soft and delicate performance style signifies a fragile naivety to the song's narrator, comforted by the permanence of the 'loving father' who remains present despite the transient nature of worldly life. Giannascoli attests to the significance of these production decisions in a recent interview for *Pitchfork*. When journalist Quinn Moreland asked whether the prominent use of modulation was a way to distance himself from the subject matter, Giannascoli offered an opposite reading: "It's more like I'm trying to capture a feeling more immediately. (. . .) I'm trying to depict the thing physically as opposed to just saying the words and hoping the listener will come around to the image" (Giannascoli, 2022a).

Although the core arrangement is that of a rock band, the edit, the mix, and the processing serve to bring harmony between this and the seemingly incongruent vocal presentation by contextualizing the track within the realm of electronic music. The drums, for example, are recognizable as a studio recording of an acoustic kit. However, the patterns are looped and quantized, there is no dynamic variation or performance embellishments (such as fills or even crashes) and the transitions between sections are marked with effects processing such as filter sweeps and bit-crushing effects.

Drum parts throughout the record are often the result of a carefully constructed collage. The self-sampling workflow is more reminiscent of the contemporary hip-hop beat-making of Kenny Segal, or the studio creations of Teo Macero and Miles Davis, than the 'faithful documentation of a performance' ethos favoured by seminal indie-rock producers such as Steve Albini or contemporaries such as Big Thief:

For God Save the Animals, Giannascoli took (. . .) to professional studios around Philadelphia, New Jersey, and New York, where engineers would record him playing his parts. They would then send him the audio stems, so Giannascoli and Portrait could mix, match and chop them up themselves.

(Ducker, 2022)

The result is a drum presentation which intentionally downplays the performative qualities of the musician in favour of studio-based processes which highlight the technologically mediated nature of the recording process. Reflecting on his own production style, Giannascoli appears to share an irreverence for the sentimentality of faithful reproduction:

As far as production, I'd imagine I have a lot in common with electronic or pop producers. I'm not trying to get something organic, I'm going in and chopping stuff up, manipulating things after the fact. I don't care about the process. I just want the product to sound exactly how I want it.

(Giannascoli, 2022a)

4.2. 'Runner'

The second track, 'Runner', is built around a conventional pop-rock arrangement although the production style contrasts with that of 'After All'. It has a more pronounced 'studio performance' feel with a comparatively natural vocal presentation. Critically, the same core instrumentation of the first track is used which helps to tie together the sound-world into a consistent syntax.

This track, along with 'Mission', 'Early Morning Waiting', 'Miracles' and 'Forgive', represent a side of the record which underplays much of the studio experimentation and high weirdness which runs through the rest of the album in favour of a production aesthetic more in keeping with the conventions of indie and folk-rock. Although it is suspected that the production process is similar to that of 'After All', the difference lies in the detail. The drums suggest a performance rather than a DAW construction; the vocals are allowed to drift in pitch and delivery; the tracks progress through changes in instrumentation and performance dynamics rather than editing and effects processing. We can envisage these tracks as representing a live performance even if in fact they are not.

Yet these tracks are rarely wholly straightforward and still contain interesting production details. One example can be found in the second verse of 'Runner' during the repeated 'couple bad things' line. Each repetition degrades in fidelity, sounding increasingly disturbed before a final cathartic scream as if expunging the guilt embedded within the lyric. It is also the first time on the record to introduce whispered vocals, a recurring theme which appears throughout multiple tracks, which in this case acts in contrast to the shouted voice to reveal a conflicted inner monologue. This performance style could be considered significant in context of the wider themes of God and religion in that it presents the lyric as if we are listening in on a private mantric prayer. The combined effect of lyric, performance, dynamic and processing serve to affirm the significance of this particular line; the narrator's prostrate cry to God for help and forgiveness.

4.3. 'S.D.O.S.'

This track is more in keeping with the production style of 'After All', once again utilizing much of the same core instrumentation, but with more prominent use of additional percussion and other textural ornamentation such as hammered dulcimer. The drums are presented similarly to the first track: dry, looped, dynamically flat and without embellishment.

This track is notable as a prominent example of some of the strangest vocal processing used on the album. The whispered first 'verse' is heavily distorted, pitch-shifted down and jittering with delay. The second verse is pitch-shifted up and heavily autotuned with individual syllables digitally stretched out to the point that conventionally undesirable artefacts, such as aliasing, are introduced. The inherent limits of the technology are exploited intentionally.

The aesthetic reappropriation of technological limitations has been traced back to the 'prepared and extended' experiments of the 20th-century avant-garde by scholars such as Caleb Kelly, who explored ideas around the 'sound of malfunction' in his 'Cracked Media' (2009). We may see parallels between the explosion of independent rock in the 1990s, via artists' use and misuse of easily accessible recording technologies, with Kelly's description of the electronic music being made within a similar time period:

> In the mid- to late 1990s experimental music witnessed an outpouring of interest as the tools of music production were transformed and rapidly expanded with the mass take-up of digital technology. The general population gained access to more affordable computer and home studio equipment, and a surge of experimentation took place. This production found its outlet in numerous small, independent record labels which released short runs of literally thousands of titles. Glitch, as the genre became known, developed as a central initial part of this outburst of creation. Producers took these newly developed, or newly accessible, musical tools and extended their use well beyond what their designers intended, pushing them until they collapsed or simply stopped working.

> *(Kelly, 2009, p. 7)*

Although the sound of 'S.D.O.S.' may appear to be rooted within the independent rock and electronic music of the 1990s, in an interview with David Folkenflik for *NPR* (2022b), Giannascoli reveals his real intention was to create a song in the same sonic palette as War's 1975 hit 'Low Rider'; a song that has become a tongue-in-cheek reference point in popular culture from *Dazed and Confused* (1993) to *The Simpsons* (2011). Giannascoli describes the process of trying to recreate something of the singular vocal style found in 'Low Rider':

> I was trying to make it low just using this pitch shifter on GarageBand, and it wasn't quite right. And then I messed around with this idea where I recorded it really fast. I sped the tempo up really fast and recorded it and then slowed the song back down to its normal tempo. And that was how I got it right, you know, and I got it to be as boomy and demonic as it sounds.
>
> *(Giannascoli, 2022b)*

Despite referencing the laid-back and playful sounds of 'Low Rider', 'S.D.O.S.' reframes these references within a darker sound-world more concerned with soul searching than the funk 'n' soul of War. The lyrics depict a tortured individual questioning their innocence, naked and curled up in the shower. The vocal processing used in each verse can be seen as symbolic of the conflicted and distorted midframe of the narrator; a split personality. The processing makes internal reference to production choices heard earlier in the record: the low whisper of the first verse once again reflects the calm quietude of an inner mantric prayer, while the childlike register of the second verse infers a vulnerability that befits the image of someone curled up in the shower questioning their own sense of morality.

4.4. 'No Bitterness'

'No Bitterness' attracted a lot of discussion from fans and critics upon its release. The oddball genre collage and loose formal structure is at once both disconcerting and amusing, sometimes verging on the absurd. Yet the combination of hook-driven contemporary pop songcraft with a disarmingly earnest lyric contains an innocent charm that listeners appear to either resonate with or recoil from.

The track starts with a 'folky' fingerpicked acoustic guitar and vocal. Giannascoli's vocal delivery is understated and unaffected, expressing a simple honesty much like the divine teacher he is singing about. But as the track progresses, we are presented with mumble-rap-like pitched-up and autotuned vocal melisma; chopped-and-screwed drums; densely syncopated rhythmic layering; banjo rolls and reversed piano swells. The song is structured unconventionally, essentially linear in form and transitions are marked by unexpected crossfades between sections. The listening experience is unmooring and unpredictable.

This sensation is compounded in the mid-section, where the track takes a marked departure of style into hyper-pop-punk. An arrangement of autotuned and distorted vocals; 'blown-out' electronic drums and 'fuzzed-out' guitars are made to be as abrasive and overwhelming as possible. Informed by the production style of popular hyper-pop tracks such as 100 Gecs' 'Money Machine' (2019), this effect is similar to the insensitively named 'ear-rape' trend in online meme culture, where recognizable tracks, often chosen for their childhood nostalgia, are bass-boosted and distorted to extreme levels for humorous and ironic effect. This willingness to draw from a wide range of serious and less-serious culture in rapid succession, even concurrently, is what led a friend of mine recently to describe listening to Alex G's music as "like scrolling down your TikTok timeline".

4.5. 'Blessing'

'Blessing' was the first track released in the run-up to *God Save The Animals*, and it appears to have confounded critics on first listen. Shaad D'Souza's coverage of the track in *Fader* described it as:

> Beginning with a blast of harsh noise and eventually opening up into a clanking industrial rock song, "Blessing" is a remarkable departure from the rest of the Alex G canon, a darkly toned dirge that makes explicit the roiling intensity that underpins so many more of Giannascoli's songs. It's a little ridiculous, but it's hard to look away—a full-blown transformation that, somehow, makes perfect sense.
>
> *(D'Souza, 2022)*

'Blessing' reportedly takes its inspiration from a much-maligned era of early-2000s radio rock and nu-metal. Giannascoli told the *New York Times* that the song comes from a period of time when he was "obsessed with Audioslave's 2002 track 'Like a Stone'" (Vanderhoof, 2022). This willingness to embrace the out-of-vogue and 'uncool' genres he grew up with is a recurrent theme in his work; the band have covered the likes of 3rd Eye Blind, Blink 182 and Coldplay. At track 8 of 13 in an album of left turns, this genre-hopping is less jarring in context, but the production takes a notedly darker turn with this track.

The heavily distorted and modulated synthesizer intro, recalling the sounds of late-2000s online micro-genre Witch-House, is abruptly displaced by a simple, metronomic rock drumbeat. Interestingly, although this song is referencing the heavy-hitting 'live' drum sounds we might associate with the records of Rick Rubin, upon first hearing this track with a friend and fellow producer I recall how we both, somewhat disparagingly, remarked that the drum sound appeared to be something akin to programmed 'stock' Logic Pro samples. The liner notes (Bandcamp, 2022) tell us this was in fact a live studio recording but the effect remains the same; as previously noted elsewhere, all dynamic and rhythmic variation is quantized out of existence, the beat is looped throughout, there are no fills or crashes and the drum beat stops with hard-cut edits. It is rock music, but with all the inherent 'liveness' purposefully removed. Once again drawing attention to the fact that this record is a digital construction and should be understood as such.

The gothic, industrial-rock sound-world extends to the whispered lead vocal, which calls to mind the hushed delivery of Trent Reznor in 'Closer' (1994) or the theatrical croon of Peter Steele with Type-O-Negative (1993). The lyric takes the form of a simple repeated stanza, here even more overtly presented as an intimate prayer to God. The hushed intimacy of the whisper is broken intermittently by a brashly grunted 'Huh', a moment of comic relief within the dark sincerity that recalls the pre-linguistic expression of machismo associated with rock vocalists from James Hetfield to Danzig.

5. Conclusions

Moylan (2020) outlines the elements of recording he defines as: spatial properties, timbre, dynamic relationships, rhythm/time, and pitch. But if we take a vocal effect like autotune, for example, the characteristic it imparts within the context of Alex G's music does not fit so neatly within these categories. There is also an aesthetic significance of such a choice in effects processing, i.e. does it and its associations appeal to the listeners' notions of good taste and authenticity within the genre?

It can also have the effect of changing the sound on the level of representation; it no longer represents the faithful capture of an indie-rock singer in a room. It removes it from the spatial representation of a material location and repositions it in that of the digital immateriality of the computer. We no longer hear the performance as existing in a real acoustic space.

By using effects processing to transform the vocal, the element which is most recognizably human in recorded music, and often revered as the primary vehicle for the interpersonal communication and expression of the singer-songwriter, Giannascoli distances the focus from the 'authentic self' and instead highlights the technologically mediated nature of the recording process. By attempting to use production to embody the sound with meaning, he treats the vocal as an instrument which can be transfigured to communicate on other levels beyond the lyrical.

The combination of multiple effects processes combine to give an overall identity to a track, one which might align with or transgress already established production styles. It is this that ultimately provides the listener with a contextual framework to make sense of it within. Listening to *God Save The Animals* gives the sense that each track is a complex web of intertextual references meant to be unpicked by the listener. Speaking to *NME*, Giannascoli has said, "I'm always true to my taste (. . .) Almost every song has a nod to something" (2022c).

The convergence of these references can be seemingly incongruous and bordering on the absurd upon first listen but given close analysis the collage reveals a carefully crafted syntax of post-internet pop culture. It is this idiosyncratic combination of reference points which has prompted many to ask whether Giannascoli is being genuine or ironic. Speaking to *Pitchfork*, he says:

> I'm basically making a collage with every song and there's no exact image I'm trying to replicate. My idea of why I did what I did changes every day. The true answer is, I don't know. Sometimes people ask if I'm being ironic or earnest, and it's like, that's not even the point. Part of my goal is to make the angle unclear, even to myself.
>
> *(Giannascoli, 2022a)*

Here we see Giannascoli reflecting on the sometimes-illogical nature of the creative process. Though it may appear that he is not primarily concerned with the pursuit of innovation, the resulting record is nonetheless replete with production choices that expand upon the conventional sonic vocabulary of contemporary indie-rock. Through the careful collage of intertextual references, styles, and contemporary production techniques, alongside a sensitive consideration for the conventions of the genre, Giannascoli is able to create a record clearly rooted within a stylistic tradition which also transcends the trend toward cultural nostalgia. The open embrace of stylistic diversity results in a music that reflects the atemporal tastes of the post-internet music consumer.

Recording analysis reveals that the surface-level reading of novelty which dominates much of the online discourse surrounding *God Save The Animals* belies an internal logic of production aesthetics which serve to enhance and deepen the symbolic value of themes explored within the domains of lyric and music. For the current recordist, an approach to production which affirms the deeper significance of production aesthetics challenges the notion of record-making as a merely technical, documentary process, while the embrace and foregrounding of digital mediation may offer a way out of the analogue mindset into new sonic territories.

References

4ad (2021) Big Thief: Certainty. Available at: https://4ad.com/news/7/9/2021/certainty (Accessed: 19/12/23)

Buchanan, I. (2010) commodity fetishism. In *A Dictionary of Critical Theory*. Oxford University Press. Retrieved 23 Aug. 2023, from www.oxfordreference.com/view/10.1093/acref/9780199532919.001.0001/acref-9780199532919-e-131

Donelson, M. (no date) Alex G Biography. Available at: www.allmusic.com/artist/alex-g-mn0003292911/biography (Accessed: 25/08/23).

D'Souza, S. (2022) Song You Need: Alex G's "Blessing" is a head dive into '90s rock. Fader. Available at: www.thefader.com/2022/05/24/song-you-need-alex-g-blessing-2022-new-music (Accessed: 01/09/23).

Ducker, E. (2022) The Search for a Unified Theory of Alex G. Available at: www.theringer.com/music/2022/9/23/23367559/alex-g-god-save-the-animals (Accessed: 25/08/23).

Giannascoli, A. (2022a) 'Alex G is Building a Mystery'. Interviewed by Quinn Moreland for Pitchfork, September 12. Available at: https://pitchfork.com/features/profile/alex-g-god-save-the-animals-interview/ (Accessed: 29/08/23).

Giannascoli, A. (2022b) Interviewed by David Folkenflik for NPR, 27 November. Available at: www.npr.org/2022/11/27/1139307799/alex-g-will-read-your-reddit-post-on-his-latest-album-god-save-the-animals (Accessed: 31/08/23).

Giannascoli, A. (2022c) Interviewed by Luke Morgan Britton for NME, 13 December. Available at: www.nme.com/features/music-interviews/alex-g-interview-2022-god-save-the-animals-3365307 (Accessed: 01/09/23).

www.unreleasedalexg.com/ (2021) (Accessed: 25/08/23).

Bandcamp. (2020) https://sandy.bandcamp.com/album/god-save-the-animals. (2022) (Accessed: 25/08/23).

Kelly, C. (2009) *Cracked media: The sound of malfunction*. Cambridge, Mass.: MIT Press.

Kholeif, O. (ed.) (2014) *You are here: Art after the internet*. Manchester: Cornerhouse.

McDermott, P. (2014) Who Is Alex G? Meet the Internet's Secret Best Songwriter. Available at: www.thefader.com/2014/05/01/who-is-alex-g-meet-the-internets-secret-best-songwriter (Accessed: 25/08/23).

Moylan, W. (2020) *Recording Analysis: How the recording shapes the song*. 1st edn. Focal Press. Available at: https://ezproxy.torontopubliclibrary.ca/login?url=https://learning.oreilly.com/library/view/~/9781317207153/?ar (Accessed: September 1, 2023).

Reynolds, S. (2012) *Retromania: Pop culture's addiction to its own past*. London: Faber.

Sartini Garner, S. (2023) Pitchfork. Available at: https://pitchfork.com/reviews/albums/unknown-mortal-orchestra-v/ (Accessed: 23/08/23).

Théberge, P. (1997) *Any sound you can imagine: Making music/consuming technology*. Hanover, NH: Wesleyan University Press, pp. 139–140.

Vanderhoof, E. (2022) Alex G, Cult Hero Songwriter, Upgrades His Sound. Vanity Fair. Available at: www.vanityfair.com/style/2022/09/alex-g-god-save-the-animals-interview (Accessed: 01/09/23).

Discography

100 Gecs (2019) 'Money Machine' *1000 Gecs*. Available at: Spotify (Accessed: 1 September 2023).

Alex G (2022) *God Save The Animals*. Available at: Spotify (Accessed: 1 September 2023)

Audioslave (2002) 'Like a Stone' *Audioslave*. Available at: Spotify (Accessed: 1 September 2023).

Big Thief (2022) *Dragon New Warm Mountain I Believe In You*. Available at: Spotify (Accessed: 1 September 2023).

Nine Inch Nails (1994) 'Closer' *The Downward Spiral*. Available at: Spotify (Accessed: 1 September 2023).

Type O Negative (1993) 'Black No. 1 (Little Miss Scare-All)' *Bloody Kisses*. Available at: Spotify (Accessed: 1 September 2023).

Weyes Blood (2022) *And In The Darkness Hearts Aglow*. Available at: Spotify (Accessed: 1 September 2023).

War (1975) 'Low Rider' *Why Can't We Be Friends*. Available at: Spotify (Accessed: 1 September 2023).

Filmography

Dazed and Confused (2003) Universal Home Video.

Sound City (2013) Gravitas. Available at: www.kanopy.com/en/mmu/video/3135678 (Accessed: September 1, 2023).

The Simpsons (2011) *A Midsummer's Nice Dream*. 20th Century Fox Television.

10

SPACE AND PLACE

Outsiders Collecting, Curating and Sharing Insider Stories and Sounds

Beth Karp

1. Introduction

The Space and Place project (Pote 2022a) investigated island belonging through the lens of community narration and field recording using a mixed-method research approach, but from the perspective of outsiderdom under the alias of Pote – Pursuit of the Extraordinary. Embedded in and working with three groups on the Isle of Lewis & Harris in the Outer Hebrides – Christian, Crofting and Harris Tweed weaving communities – a principal aim of the project was to explore narrative and sound, utilising community narration and contextual field recording to aid the process of storytelling.

In this chapter the relevant literature will be reviewed, methodology and methods discussed followed by personal reflections, 'behind the sound' accounts and the author's observations, before the concluding summary.

2. Literature

The methodological framework for this project was informed and shaped by several literature sources, centred around field recording, narrative and storytelling, narrative inquiry, self-reflexive narrative, community narration and cultural identity.

3. Narrative and Storytelling

Due to the desire to understand the craft of storytelling in its aural form, it was pertinent to investigate narrative and storytelling through music. Tyler Caldas (2020) shares how the nature of storytelling is everywhere, and that humans are hardwired to seek out stories. We have long communicated through stories, and the telling of our own via various means can be seen throughout history. Music is no different. "Simply put, a narrative is a story. And if there is one thing the human brain is innately wired to do (and do exceptionally well), it is to seek out and understand stories" (np).

Referring to the works of James Andean (2014) and his studies of acousmatic music, the author understood their own practice went against the interpretations of musique concrete from Pierre Schaeffer's standpoint. Whereby, the characteristics of a sound are central to their use, with the significative identity of the sounds removed (p.1). The intention for this project was to utilise and

DOI: 10.4324/9781003396550-10

highlight these signifiers and their identities, and, with careful consideration, alongside the musical potential of each sound, create a piece that rejoices in its creation, its narrative and becoming.

Andean unpacks the notions of listeners, even unconsciously, conjuring mental images built from the role of the sound as a signifier, but goes on to explain:

> The unique beauty of the genre lies precisely in this duality: the purely musical world on the one hand, where the sounds are composed and appreciated for their musical or sonic properties and on the other hand, the stream of sources and imagined gestures that these sounds evoke. (p.1)

This was important to approaching musical elements of the pieces, although not acousmatic music due to the aural content. The desired effect was for the music to encapsulate the environments and climates (social and literal) of these peoples' lives to help convey the realness of their stories. In essence, to make each piece an immersive experience, drawing on the sounds of each community, the music elements and textural sounds in themselves would tell the story alongside the voices.

Similarly, Tullis Rennie's (2014) discourse on socio-sonic electroacoustic composition discusses how

> Electroacoustic composers, in their broadest terms, have regularly considered sounds simultaneously as contextualised and as sound-object, but most often in ways in which the final composition errs on the side of the abstract. Works that are 'inspired by a visit to' or 'using sounds collected in place x or scenario y' rarely demonstrate an equal relevance to the inherent social context of the source sounds alongside their intrinsic sonic qualities.
>
> *(p.117)*

This was part of the intention for the Space and Place project and helped make sense of this researcher's ideas that whilst the pieces are abstract, they embody the social context of the source sounds fully, including their sonic qualities. This was a key component of the approach, laying aside personal tropes and preferences, creating an entirely new creative practice for the artist, to give space to share the social and cultural aspects investigated.

Drawing on findings from an earlier project (2021) exploring field recording, self-reflexive narrative and sound manipulation to familiarise the researcher with the processes required to undertake work of this nature, it was a hope to evolve personal processes and practice to sit outside the artists' normal creative boundaries. The research further explored these ideas of found sounds, alongside oral accounts to share and build narrative where normally the inclination would be to write lyrics.

4. Self-reflexive Narrative

In the aforementioned previous works drawing on the ideas of self-reflexive narrative (Karp, 2021), this researcher found the approach from Tullis Rennie and Isobel Anderson to be a turning point in their own practice, enabling them to move beyond the confines of what was known, and of how to get from idea to completed piece, into an arena of unknowns and vast possibilities (2016, 2021).

Anderson and Rennie (2016), looking at self-reflexive narrative and field recording, drew on the work of others, arguing:

> that field recordings of this kind are 'autotopographic' (Heddon 2008) in nature: narrating both self through site and site through self, within the medium of sound. For Elinor Ochs and Lisa

Capps, narrative and self are inextricably linked; they state that 'narrative is simultaneously born out of experience and gives shape to experience' (Ochs and Capps, 1996:19).

(p.223)

Relating this to the exploration into the communities, it became as much about discovering a sense of personal belonging as uncovering theirs. Furthermore, taking the perspective that the subjects being interviewed were sharing their story, narrating their experiences and sense of self in terms of space and place took precedence over the sonic outcomes.

5. Narrative Inquiry

The initial research question was: How do we carry out an inquiry of narrative in the context of making music utilising field recording and self-reflexive narrative? Understanding the approaches came first, taking note to comprehend the autobiographical nature of each subject's account during interviews; drawing upon narrative memory, community narration and the charting of a life story's path. This was followed by considering the compositional approach to the narrative inquiry.

The value of autobiographical narrative inquiry offers an

inroad into exploring the dynamic features of narrative inquiry as applied to the study of lives. It can help show how and why narrative inquiry might lessen the distance between science and art and thereby open the way toward a more integrated, adequate, and humane vision for studying the human realm.

(Freeman, 2007, p.120).

As the purpose of this project was to engage with three different communities, the peoples of said communities would share autobiographical accounts. Each would, undoubtedly vary, but offer a window into the lives of the people in a musicological context.

It was interesting to uncover any shared identity points in the narrative as Freeman explains:

There are, for instance, cultures where human personhood is framed less in terms of an individual identity, with its unique and unrepeatable story of coming-to-be, than in terms of its social place, its role in a cultural pattern that may be deemed timeless. Narrative remains a relevant category in such cultures but may be considered more a matter of public, rather than private, property.

(Freeman, 2007, p.121)

Narrative memory is "the subset of episodic and semantic memory in a cognitive storing information that presents narrative features" (León, 2016). It's said to be found everywhere and "According to some models, this is due to the hypothesis that narrative is not only a successful way of communication, but a specific way of structuring knowledge" (p.1).

The simple nature of how we draw on memory and understand narrative as a species enables us to connect with story with exceptional ease. This notion was helpful when approaching a study based on telling stories, sharing and building narratives from lived perspectives, both individual and communal, and leading us to community narration.

6. Community Narration

Due to the nature of the groups interviewed and the structure of the project, it was important to understand more about community narration and how or why the stories shared by people in a community might be affected or shaped by being part of said community.

Community narration is the comprehension of the culture of a group or community, taking into account its "belief systems, values, social norms, and practices" (Olson & Jason, 2011).

The goal of Community Narration is to understand the identity or identities of an organization or community. The group's identity, signified by the bipolar constructs, is the lens through which the members see significant parts of their world, reflecting how the community narrates its shared story.

(Olson & Jason, 2011).

Stapleton and Wilson open their paper on community narrative by summarising "Narrative theory distinguishes between the theme of a story and its form or 'telling'" (Stapleton & Wilson, 2017).

Rappaport (2000) distinguishes between personal stories and community narratives, associating the story to the individual and the narrative to the community, noting that stories and narratives are intimately tied with one another. Each community has a unique set of narratives that are a source of growth, and a way for a community to creatively find its alternative narratives, which are contrasted with other dominant narratives in society. The contexts and cognitive constructs associated with such alternative and dominant narratives are critical to understanding the relationship or fit between an individual and the larger organization or community.

(Olson & Jason, 2011, np)

It was of particular interest to see if subjects in the shared cultures identified some shared narratives and if so, perhaps of more interest would be the way that informed the composition, which we'll discuss shortly.

7. Cultural Identity – People and Place

Màiri Robertson explored the ideas of how relationships to the past are constructed and the impact of this on the future of cultural identity, in the context of culture, environment and intangible heritage of the Western Isles (2009).
    Robertson draws on

relationships between sense of place, identity and possession and whether in the case of the Gàidheal-tachd (in both Scotland and Ireland), attachments to place are intrinsic to identity, rather than to buildings or monuments. Periods of dispossession and being psychically absent from the landscape at certain points in the past at community, collective and individual levels have influence upon 'sense of place'."

(Robertson, 2009)

A basic distinction is drawn between general recognition of certain areas as places and a personal sense of place, both of which we all live with intimately. Thus, memories of place may depend, to some degree, upon landscape, but the sense of home as a place may be grounded more in human relationships than solely upon the memory of a landscape (Robertson, 2009). This idea of place was key to the project, for it was this sense of place that was being captured. "Emerging during the 1970s, human geography emphasised ideas of 'place' and its connection with self" (Anderson & Rennie, 2016, p.225). Our connection to a place is not to be overlooked, it clearly forms a building block in the formation of identity and a sense of belonging, which in turn builds narrative.

Stories play a role in our personal narrative formation and identity from when we are children onwards.

Memory plays an important part of identity formation and creating a positive sense of self. As a child develops and has experiences, there is a part of the brain that creates a story from these experiences and over time there is a sense of self that develops. This is known as Autobiographical Memory.

(Hull, 2018)

Which, Szilas argues, "happens with narrative-related cognitive processes (emotions or chronological ordering)" (Leon, 2016). Indicating how our personal narratives are built from a young developmental age, that we learn to formulate stories in an effort to remember. Recalling these stories brings the past back to life and can be as simple as talking about the memory, looking at a photo or hearing a piece of music. This idea of memories being triggered has long been assessed and discussed. Cathy Lane and Nye Parry draws on the works of Voegelin, who believes memory isn't something we conjure from "the past . . . but that is triggered by the current event and becomes materialised in the now" (2006). Exploring cultural identity further leads us to understand the nature of being human and how we relate to our environment, our place.

Experience is the organic intertwining of living human beings and their natural and built environment. For Dewey (1934), human beings are not "subjects" or "isolated individuals" who have to build bridges to go over to other human beings or the things of nature; human beings are originally and continually tied to their environment, organically related to it, changing it even as it changes them.

(Bach, 2007)

8. Methodology

Using mixed-method research (MMR) approaches enabled the examination of the culture, heritage and future hopes of the people of the island and created sonic art that reflected the nature of these communities, it also afforded the opportunity to examine and understand the positionality of outsiderness, belonging and creative practice development.

The MMR included inductive and qualitative research methods underpinned by interpretivist and constructionist philosophies, which then drew on a rich in-depth ethnographic study, literature review, archival and desk-based research, autoethnographic analysis for the exploration of my own outsiderness and sense of belonging for the outsider piece, and phenomenology utilising thematic and content analysis for organising and categorising narrative elements from interviews with locals.

9. Method

As a practitioner-researcher using MMR approaches, it is argued that due to its new emerging, and therefore somewhat incommensurable framework, the researcher can mix worldview paradigms (Ghiara, 2020). The value of using MMR offered engagement in a variety of practices within one project yet formed cohesive outcomes that were underpinned and rigorous.

In this case, the research drew on the strategies of ethnography, phenomenology and a small amount of archival research as the first point of data collection, followed by a multi-method data analysis process for listening through the recorded audio: using both content and thematic analysis.

"TA is a method for systematically identifying, organising, and offering insight into, patterns of meaning (themes) across a dataset" (Braun & Clarke, 2012). The use of TA gave the building blocks for forming the narrative from the multiple oral accounts from the participants in a methodical adaptable format.

The project required a longitudinal timeline to accommodate the constraints of data collection; including multiple data sources and their availability, and Covid restrictions.

The data collected were the stories and experiences of people from each set community group. Using semi-structured interviews over Zoom and in person enabled the gathering of data on the themes of present, past and future within the context of each community – i.e. church, weaving and crofting. Kvale and Brinkmann (2009) stated on the craft of quantitative research interviewing: "The research interview is based on the conversations of daily life . . . where knowledge is constructed in the inter-action between the interviewer and the interviewee" (p.2). The correlation with the constructionist paradigm and how, as interviewer, the researcher is constructing the knowledge in conjunction with the interviewees, gave a clearer insight into the approach and method for interviewing, further highlighting the importance of not imprinting personal interpretations, where possible.

The sampling was somewhat defined by each group, with an intentional cross-section of participants to gain a comparative representation of each group – a purposive sampling. So, a mixture of indigenous islanders and those who had moved to the island and chose to dwell and be a part of the community there. There, however, were also elements of random selection and convenience to the sampling. Interviews were recorded in-person and online via Zoom to fit ethical considerations – impacted further by the Covid-19 pandemic. For in-person interviews, a combination of mobile phone and Zoom H4n Pro microphones were used to capture audio. Similarly, for field recording, both microphones were used, alongside the addition of contact microphones for sonic and tonal variations. Audio captured was then uploaded into separate applicable Logic Pro projects and analysed utilising the forementioned methods. The musical components were composed and created using both qualitative and quantitative data.

10.   Outsider/Insider

As part of the research explored island belonging in the Outer Hebrides, for me as an outsider English woman, my approach included autoethnographic accounts, observations and reflective/ reflexive songwriting in response to said outsiderness. Therefore, we will move briefly into autoethnographic foregrounding.

11.   Foregrounding as Outsider

Having relocated to the Isle of Lewis & Harris by choice and not for the purpose of research, I began embedding in island life. I had not visited the island before relocating, but from the moment of our arrival, I felt connected, tapped in somehow. Driving through the wild landscape, encountering changeable weather around each corner of the road carved into the hillsides and bog, I felt drawn to the place, its shape and form. The tranquil calmness the island appeared to embody, ran through the pockets of communities spread in small clusters around the island, like it was spread through the air, the sound of the tide, hills and the peat-soaked smoke rising from small croft house roofs. I wanted to better understand this strange sense of belonging I felt to the place and its people. I was intrigued by the sounds of everyday life that surrounded me; the simplicity and calm of this Hebridean haven could almost be summed up in its sonic nature. By exploring field recording,

I became entangled and inspired by the possibilities to capture, celebrate and uncover the stories of this island and its people, and I hoped to extrapolate my own sense of belonging alongside the natives and other outsiders who called this place home.

There is a term that came up a lot in my research into these island communities on the Isle of Lewis & Harris; it is a Gaelic word to describe the sense of belongingness the island folk seem to have here. 'Buntanas' means more than belonging, it is about your connection to the land, the place, the people of that place, the heritage, being of the land and it remaining with you, and then when you die you become part of the land here. The contextual music made to sit alongside the collective community narrative began all as sounds of the island, the people, their jobs and roles of everyday life. These sounds all form the body of these communities' lives and are as deeply rooted in their belonging as their heritage.

12. Behind the Sound

The notion of me as an 'Outsider collecting Insider stories' swiftly evolved into a new realm of the outsider curating, creating and sharing insider stories. The ethical implications of this and the ensuing power shift were mitigated throughout the process and approach of the research being deeply embedded in a purposeful pursuit and intention to uncover and share the contextual narratives and sounds of the communities in a way that did not draw on personal tropes or preferences.

In this bid to capture the true nature of islandness and belonging amongst the three communities, the insider stories collected through semi-structured interviews, not only formed the narrative of the pieces but also informed the music/sound elements that lay alongside the stories.

The intention was to capture the true nature of the island and communities in the musical elements created using the field recordings, this meant not just stopping at simply throwing the sounds in to reinforce the narrative elements being shared, but instead exploring the culture of the stories and building 'musical' elements that had that culture embedded in them.

Every piece had a nod to the heritage of traditional highland music using a highland bagpipe drone sound, created using transient peaks in an EQ tuned to harmonics of B flat from the captured field recordings relevant to each community and then used in various ways in each piece.

The church community was tackled first, which naturally had more musical elements due to the nature of this community's worship and Gaelic psalm singing on the island. So, alongside these natural elements, a psalm-sung melody was created utilising a synthesised sample collected from the community to play at a specific point in the shared narrative. At later intervals, real Gaelic psalm singing is heard, then in the future section, drawing on the developing Christian churches on the island, an island youth worship band singing what would be deemed as modern worship there is used to share the evolving of the 'church' on the island.

In the crofting piece, this researcher utilised four maps relating to land agricultural viability for the island that displayed factors impacting crofting (soil, land, wind exposure and precipitation data for the island), not wanting to just sonify the data but integrating the shape and form of the island itself to become embedded in and forming the music. This was carried out by drawing the map of the island using MIDI art, and using the data from the maps combined with contextual natural element sounds recorded on the island such as wind, rain and boggy footprints, etc. The data was translated into synthesised samples to make the musical elements of this piece [Figure 10.1].

Several methods were used to input the island's form into the piece, including panning patterns and step-sequenced beats – although it doesn't sound the prettiest in terms of its sonic properties, the representative nature of these elements was of greater importance compositionally and philosophically than the sonic outcome.

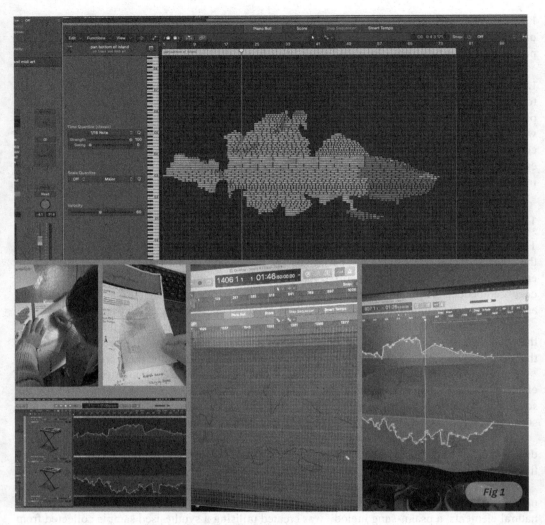

FIGURE 10.1 Map of the island in completed MIDI art form. Tracing maps into sequenced steps. Panning patterns to form island outlines.

For the weaving community (Harris Tweed) piece, the information of the weavers' pattern stamp cards, alongside the tweed colour data were translated to create melody patterns for each section of the piece – present, past and future. For these data-sonified sections, era-specific looms used on the island at different points in this industry were recorded and utilised to represent the time periods being examined. This, alongside the correlating patterns of Harris Tweed and its era-specific colours, formed the melodic backdrop to the narrative shared by the weaving community.

To create the future section for this piece the sound was created still using the rapier loom (which is a modern double-width loom) – as the definition of Harris Tweed states that it is hand woven so it will not be robotic machines creating Harris Tweed. Therefore it made sense to use the latest looms used on the island as a backdrop and bring in the future sound through what was shared in the narrative. If you listen to the whole piece you will hear how "The Harris Tweed industry is aiming to attract more in the area of ladies wear fashion" (Harris Tweed representative

FIGURE 10.2   Recording Rapier double-width loom. Harris Tweed pattern stamp cards and corresponding pattern decoding with hex codes of Harris Tweed colours. Crofter/ Weaver Morris explains how the stamp card works. Single-width loom as used 100 years ago.

Lorna, 2022), so this is where the creative expression was aimed – selecting a complex weaving pattern and then choosing vivid and wild colours to depict the future narrative. The colours are all taken from the Hebrides Tweed website, the hexadecimal codes representing each colour were then applied to the pattern stamp card for the six-colour check [Figure 10.2].

By painstakingly translating these into the MIDI roll, using a synth created with samples including ambient EQing and stretched loom samples, the melodic backdrop to this future section was crafted. It has a robotic futuristic quality, created by tweaking effect/modulation/LFO/ oscillation and filter parameters in the multi-sample synth.

For the outsider piece, as my autoethnographic exploration of the self in the context of outsiderdom on the island, the author drew on their traditional creative practice as a songwriter, sharing observations and longing in the lyrics, sung along to acoustic guitar, purposely simplistic in its melodic nature to aid the evocativeness of its sonic backdrop and to aid the storytelling process. It carefully and contextually shared the identity of the self – their own space and place alongside their position of outsider amongst the communities they were 'in'. This piece enabled the researcher to examine their own identity and the connection between personal narratives and the forming of an evolving identity from birth to a proposed future in that outsider position. For example, the song's structure follows the pattern of present, past and future, as per the research interrogation (Figure 10.3).

13.   Thinking Outside the Box

From a position of outsiderdom, a fly-on-the-wall view, and perhaps an off-island upbringing, shaped the author's view. In that, there were displays of oppression, sexism and gender inequality as a symptom of the culture on the island. It appeared to sit quietly at the foundation of life here – like a natural system of their society and culture.

There was a simplicity to the desire to live by many indigenous island folk. A female neighbour referred to herself as 'basic' and described the drive to do more or be more simply as not something they wanted – they like their lives, working in standard roles, their local strolls, seeing family and looking after them. She drew a comparison between the pursuit of the research and my personal long-term goal as being less basic than their hopes, referring to herself and her friends (Kirsty, 2022). Others spoke of societies on the mainland being more concerned with 'keeping up with the Joneses' and in a constant state to outdo one another. The people of this place seek to just live their lives (Morris, 2022).

Alternatively, a difference was noticeable in island dwellers who had adopted the island as their home. An outward enthusiasm was noted in the outsiders who had relocated to the island. They shared what they did and why with an eager passion and pride, whereas the locals doing the equivalent roles in their communities didn't see what they were doing with the same notion. Often dispelling the validity to share with me about what they did in their daily lives, not seeing it as anything special, but moreover believing it was 'just what I do', nothing to write home about or to outwardly show pride in the successes of.

The heritage of islandness and community shows a richness in roles and the way these communities functioned was greatly dependent on the gendered roles – men went out to sea and women stayed at home looking after the family – but also kept the croft going – tending the animals and land of the croft, their children and the surrounding village community.

The men would return, bring much-needed resources and reserves, and often weave tweed. The cycle worked well. Village communities thrived, children thrived and grew up passionate about the land and sea, their heritage and their connection to the place. It was embedded and ingrained in their being. A deep sense of satisfaction seemed to sit within them even though it was often grueling work in a myriad of poor weather conditions.

The schools were filled, and children grew up taking part in community – working as part of a well-oiled machine – they learnt to help because it was the done thing, a natural role and way to be part of the communities they dwelled in, they understood what it was to care for and look out for one another (Morag, 2022).

The roles of women here included physically demanding duties – they cut peats, moved sheep, slaughtered chickens, carrying the weight of generations, caring for not only their own family but that of the needs of the communities around them. They were vital to its survival but also to its joy and togetherness.

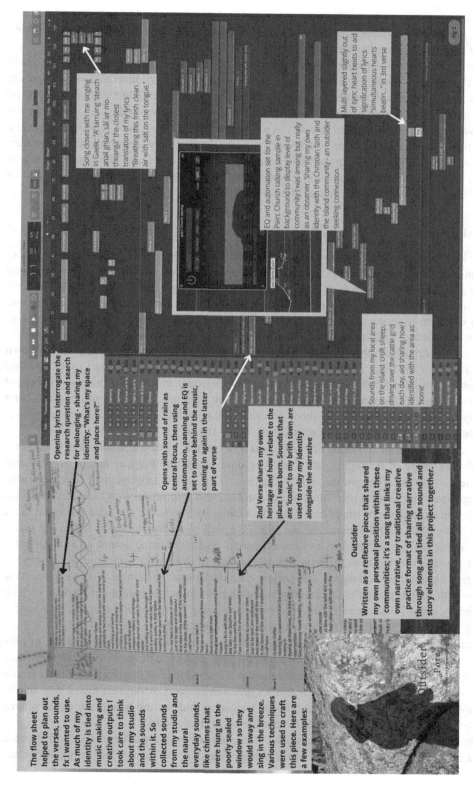

FIGURE 10.3   Outsider breakdown – sharing the links between identity, narrative and sound processes for this piece.

Did these women equate themselves as unequal to men? They appeared to understand what was needed and got on with it, happily, often with a quiet reserved pride, ensuring everything was done well, rather than questioning the status quo here.

Caroline Criado-Perez (2019) discusses the data gap of female representation across the world of statistical and factual policy making to have contributed to the corruption of how women see themselves and, in the past, have been trapped in a myth that women couldn't work in male roles. "The persistence of this myth continues to affect how we see ourselves today . . . and how we see ourselves is not a minor concern. Identity is a potent force that we ignore at our peril" (p.21). Further to this, the feminist data gap is discussed within island society from research undertaken on the Faroe Isles: "a research gap exists in terms of taking an explicitly feminist approach to studying gender and social inequalities in island settings while deliberating islandness as part of the intersectional analysis" (Gaini, Pristed Nielsen, 2020).

14. Conclusions

One of the key elements of the Space and Place project was to understand island belonging from the position of my own outsiderdom. The more I researched the communities, the more I realised how I was searching for the sense of belonging I was witnessing here. I draw back to the term 'buntanas' and its deeply rooted sense of belongingness, that one is born into, and cannot otherwise gain as an outsider. It appears that each person's life within this belonged community holds so much more significance and reverence for the people, place and connection to the environment. I do not know how to obtain this level of belonging personally and do not feel such with my own ancestory, family or the place in which I was born and grew up. The opportunity to seek and find out more about the island communities really gave me personal insight into the reasons behind wanting to feel a sense of connection or to understand why I felt connected to a place where I had no ties. "By looking into the soul of another we often find ourselves delving just as deep into our own private worlds of identity and place" (Cowell, 1997: 46).

I discovered that as an outsider, not only was it expected by some locals that I wouldn't understand their sense of belonging, but that I could never capture the true nature of what that belongingness – 'buntanas' meant through my research. One of the participants, Frank Rennie of Galson, explained this term to me: "they are of the land and will return to it" (2022). Although I had a different experience of island belonging than those who were 'of' the island, I found that I felt drawn to the place and the people of the place. Regardless of how knitted I felt to the place, I was not of the place – the project explored this and captured lasting sound art that shares a wealth of rich stories of the lives, heritage and islandness. It also encapsulates the sense of being an outsider, in a pursuit for belonging. The project extended further to helping others explore their own stories with sounds using field recording, narrative and songwriting and I ran workshops with a local primary school that created their own piece exploring narrative and identity as a school community – their own Space and Place (Sgoil Na Pairc Anthem (Pote, 2022b)).

The creative practices adopted and explored in this project formed an understanding of new ways to utilise field recording in my music making but also created an entirely new creative practice for me in making and thinking about music. It embedded me in mixed-method research (MMR) practices and sparked intrigue for further research projects. Alongside this, I was able to examine and formulate new ways of drawing on, triggering and sharing memories, using sounds and sonic effects to tie narrative elements to tangible contextual sound.

I had anticipated collecting and sharing stories alongside exploring creative practice and research but gained so much more in my pursuit of the extraordinary on this island. Not only did I create lasting sound art that encapsulated this wonderful space and place, its people, heritage and culture but also gleaned more of an understanding of what it means to belong to a place, in connection and a way of living that perhaps many of us have lost along the way. There was an exceptional beauty to life in the Outer Hebrides, which is intertwined with its present, past and future. As it turned out, collecting insider stories and sounds as an outsider wasn't as controversial as it might initially seem. I believe the perspectives outsiders can bring to rooted cultures are interesting and perhaps enables new perspectives, and modes of conducting, sharing and transmitting thoughts/stories/ music/art etc. in new ways that can add to the richness of those cultures.

References

Andean, J. (2014). Sound and narrative: Acousmatic composition as artistic research, Journal of Sonic Studies, 07 (website), available online at www.researchcatalogue.net/view/86118/86119/3622/0

Anderson, I., & Rennie, T. (2021). New Directions in Field Recording: Pluralising the Field Seminar – MeCCSA Sound Studies network. Available online at www.youtube.com/watch?v=y-dV9z0BK0U

Anderson, I., & Rennie, T. (2016). Thoughts in the field: Self-reflexive narrative in field recording, *Organised Sound*, 21(3), pp. 222–232.

Bach, H. (2007). Composing a Visual Narrative Inquiry. In D. Jean Clandinin (Ed.) *Handbook of Narrative Inquiry: Mapping a Methodology*. Thousand Oaks, CA: SAGE, pp. 280–307. Available online at: www. doi.org/10.4135/9781452226552

Braun, V., & Clarke, V. (2012). Thematic analysis. In H. Cooper, P. M. Camic, D. L. Long, A. T. Panter, D. Rindskopf, K. J. Sher (Eds.) APA handbook of research methods in psychology, Vol. 2: Research designs: Quantitative, qualitative, neuropsychological, and biological (pp.57–71). www.researchgate.net/ publication/269930410_Thematic_analysis

Caldas, T. (2020). How Music Crafts A Narrative – Everything Is Noise. Everything Is Noise. Available online at: https://everythingisnoise.net/features/how-music-crafts-a-narrative/

Cowell, A. (1997) Defining and Expressing a Valid Personal Reality, (unpublished research report), University of Southern Queensland.

Criado-Perez, C. (2019). *Invisible Women*. Penguin.

Freeman, M. (2007). Autobiographical understanding and narrative inquiry. In: *Handbook of Narrative Inquiry: Mapping a Methodology*. D. Jean Clandinin (Ed.) Thousand Oaks, CA: SAGE, pp. 120–145. Available at: www.doi.org/10.4135/9781452226552

Ghiara, V. (2020). 'Disambiguating the role of paradigms in mixed methods research', *Journal of Mixed Methods Research*, 14(1), pp. 11–25. doi: 10.1177/1558689818819928

Gaini, F., & Pristed Nielsen, H. (2020). *Gender and Island Communities. Routledge.* Available online at taylorfrancis.com

Hull, K. (2018). The Crucial Role of Memory in Creating Identity and a Positive Sense of Self. [online]. Available online at: https://drkevinhull.com/blog/2018/4/21/the-crucial-role-of-memory-in-creating-ident ity-and-a-positive-sense-of-self

Karp, B. (2021). An exploration of how, as a composer, I can utilise field recording creatively. Unpublished. Works available at: Pote explores field recording and Self-Reflexive Narrative | Pote (pursuitextraordinary. wixsite.com).

Kvale, S., & Brinkmann, S. (2009). *InterViews: Learning the Craft of Qualitative Research Interviewing*. Sage. Available online at: https://books.google.co.uk/books?hl=en&lr=&id=bZGvwsP1BRwC&oi= fnd&pg=PR1&dq=Kvale,+S.,+and+Brinkmann,+S.+(2009).+InterViews:+Learning+the+craft+of+qual itative+research+interviewing.+Los+Angeles:+Sage.&ots=q8HSwhszKh&sig=zyQeVpDK_idCKVuc McQFCa_z5QQ#v=onepage&q&f=false

Lane, C., & Parry, N. (2006). Sound, history and memory' thematic issue, *Organised Sound*, 11(1), pp. 1–2. doi: 10.1017/S135577180600001X

León, C. (2016). An architecture of narrative memory, *Biologically Inspired Cognitive Architectures*, 16, pp. 19–33. www.sciencedirect.com/science/article/pii/S2212683X16300184

Olson, B. D., & Jason, L. A. (2011). The Community Narration (CN) approach: Understanding a group's identity and cognitive constructs through personal and community narratives, *Global Journal of Community Psychology Practice*, [online] Available online at: www.ncbi.nlm.nih.gov/pmc/articles/PMC 3821795/ [Accessed 26 December 2021].

Rennie, T. (2014). Socio-Sonic: An ethnographic methodology for electroacoustic composition, *Organised Sound*, 19(2), pp. 117–124. www.doi.org/10.1017/S1355771814000053

Robertson, M. (2009). Àite Dachaidh: Re-connecting People with Place—Island Landscapes and Intangible Heritage, *International Journal of Heritage Studies*, 15(2-3), 153–162, www.doi.org/10.1080/135272 50902890639

Stapleton, K., & Wilson, J. (2017). Telling the story: Meaning making in a community narrative, *Journal of Pragmatics*, 108, pp. 60–80 Accessed December 2021 www.sciencedirect.com/science/article/pii/S03782 16616306233

Interviews references

Kirsty. (2022). Interview with Beth Karp, February, Lewis.

Frank. (2022). Interview with Beth Karp, March, Lewis.

Lorna. (2022). Interview with Beth Karp, April, Lewis.

Morris. (2022). Interview with Beth Karp, May, Lewis.

Morag. (2022). Interview with Beth Karp, May, Lewis.

Discography

Pote (2022a) The Space & Place Project. [Digital] Available at: https://pursuitextraordinary.wixsite.com/pote/ pursuit-of-the-extraordinary-pote-embarking-on-vast-project-of-space-place or https://pursuitoftheextrao rdinary.bandcamp.com/album/the-space-place-project

Pote (2022b) Sgoil Na Pairc Anthem – Space & Place. Available with corresponding collaborative video made with Jan Schouten: https://youtu.be/ahJwVaIh_oQ?feature=shared

Suggested further exploration: www.tobarandualchais.co.uk/

11

AURAL ARCHITECTURE

Integrating Site into Composition

Bob Birch

1. Introduction

Composed music does not usually consider the space in which it will be performed. Purpose-built venues are often designed to be a 'blank canvas' to reduce the amount of acoustic interference to the intended work. Anyone who has performed live music, particularly in sites where musical performance is not the intended function, will know how the acoustic environment causes complex changes to the perceived sound; changes that are usually not welcomed. For example, the reverberation time will have an impact on the articulation or choice of tempo and may even mean some music is unsuited to the acoustic environment. The composer and performer can be inspired by or, similarly, limited by the acoustic characteristics of the performance space. It is not just that the sound is transformed by the space, the relationship is reciprocal: the space is transformed by the sound. The environment influences the perception of the sound and simultaneously the sound alters the perception of the environment. The building's response, however, is 'flawed'. Physical spaces act as filters, removing or enhancing particular frequencies through reverberation and resonance. Yet these flaws "establish the character of an environment" (Gendreau 2011, p.33). The act of integrating the site is to identify these flaws, which can then become features, "elements of the individual language of the structure" (ibid., p.40). By projecting sound where it is not expected, or placed in relation to architectural objects, it becomes recontextualised and this allows sound "to carry extra musical connotations" (Sambolec 2011, p.55). This paper discusses ways of integrating site into musical composition. Drawing on the ideas of Blesser and Salter (2007, p.5), aural architecture is a term used to describe the "properties of a space that can be experienced by listening". My practice involves composing music that highlights the acoustic qualities of the space in which the music plays out, exposing and integrating the aural architecture. I often use irregular multichannel speaker arrays; in two pieces, this necessitated the development of a software-hardware solution that uses repurposed joysticks. In one work, this enabled eight performers to interact with and create the composition in real time, by investigating the site using sound through sixteen carefully placed loudspeakers. The same hardware control was employed to allow an imagined aural architecture to be created within a small telephone isolation booth. In another work, I used acoustic sources in the form of mechanical music boxes. These simple devices were augmented by the addition of small electric motors and motion sensors. In this interactive work, the movement of visitors within the performance space creates the piece.

DOI: 10.4324/9781003396550-11

My practice research is inherently about modifying or misusing existing technology to create a technical solution for a creative output. Working with site necessitates the development of audio tools or technologies, to "make something . . . in order to make something" (Young 2016, p.154). The purpose of this paper is to discuss the development of my practice research through these works to show the potential in integrating site with composition. Using empirical methods, I have found a rich and rewarding outcome where sound and space become interconnected through composition. Spaces with challenging acoustic characteristics can be celebrated rather than overlooked.

2.   Background

During the modernist era, architecture was thought of as a purely spatial and visual form, whilst music was considered to be purely temporal (Till 2016; Sheridan & Lengen 2003). Le Corbusier epitomises the modernist view: "architecture is in space, in extent, in depth, in height: it is volumes and circulation" (cited in Till 2016, p.165). In fact, music is inherently spatial as well as temporal. Spatial hearing is, according to Blauert (2008), a tautology; hearing is always spatial due to the way our auditory system experiences sound. It is performed in space and thus is shaped by the position of the sound sources and is affected by the acoustic properties of the space in which it is heard. Architecture is also not experienced in a purely visual sense but "through the senses of sound, touch, and smell, the modalities of time, association, and memory, and the contingencies of social use" (Till 2016, p.166). Architect Daniel Libeskind suggests: "Buildings provide spaces for living but are also de facto instruments, giving shape to the sound of the world" (Lutz 2007, p.172).

The properties of space can be experienced by purely aural rather than by visual senses, in what Blesser and Salter (2007) call aural architecture. As multiple sound sources interact with the various surfaces, objects and geometries of a space, a complex aural environment is created. It is the human experience of this environment that gives a space its aural qualities, or 'personality', which can affect our moods and emotions. All spaces have an aural architecture, built and natural environments alike. Through sensing changes in pressure, the human auditory system is able to detect both complex and subtle features of the environment – to "sense space by listening" (ibid., p.1). The phenomenology of aural space is complex; it is a combination of spatial and auditory perception and also personal and cultural values. At one end of the process there is little difference between the individual but at the other, sounds have personal meanings and associations. It is the stages of sensation (detection), perception (recognition), and affect (meaning) that Blesser and Salter (2007) suggest make up 'auditory spatial awareness'. It has a profound influence on social behaviour, encouraging quiet and reflective listening (for example, in a church), or feelings of isolation or cohesion. The auditory system's ability to sense the space we are in can also provide navigational cues: we can 'see' with our ears. Our auditory system is working subconsciously, detecting characteristics of the space we are in, "decoding of the 'feeling' of the room we are in" (Emmerson 1998, p.137). Aural space also has an aesthetic value – a space without acoustic features is as "sterile and boring as barren, grey walls" (Blesser & Salter 2007, p.11). Composer Christine McCombe (2001, p.3) asserts: "The way we experience a space is determined largely by our aural perception and our physical presence in that space." Auditory spatial awareness can also enhance the experience of music and in this case the space becomes an extension of the musical performance.

A site-specific composer must consider both the physical acoustics (the spatial acoustics which describe the physical properties of sound waves within the space) and the aural architecture (the composers' intention for how the audience experiences the space). The space should therefore be considered an intrinsic part of the instruments, and a tool to be used by the composer. In the same way that light is used to illuminate visual architecture, sonic events can be used to 'illuminate' the site – to emphasise the unique characteristics and augment the listeners' experience of the aural architecture. Blesser & Salter (2007) suggest that in order to have any meaningful aural awareness, the listener must engage in attentive (or active) listening. Sound can be heard by the individual, but can (and should) be experienced as a "sonic event and [its] modification by the aural architecture" (ibid., p.15). Hearing is the detection of sound, but listening is the "active attention or reaction to the meaning, emotions, and the symbolism contained within sound" (ibid., p.5). Although it is not possible to force visitors to listen, it is my intention through the presentation of sound works to encourage an active listening behaviour, a greater appreciation of aural architecture and improved auditory spatial awareness. The individual's unique experience of aural space cannot be controlled by the composer but is dependent on sonic events: to reveal the aural details of a site. An understanding of the propagation and interaction of sound within the space can be used by the site-specific composer to transform the sound in response to the architectural features. Architectural objects such as archways or vaulted ceilings form part of the visual experience and almost all visual embellishments have an acoustic effect: they change the sound around them. Blesser and Salter (2007, p.52) call these "aural embellishments". Whether large or small, they produce changes in the acoustic attributes by transforming the sound events, adding richness and variation to the aural experience. As these aural embellishments create aural variety, they are not usually desirable in traditional musical spaces. A concert hall is ideally acoustically uniform to allow a consistent experience for the paying audience (Baranek 1996). However, aural embellishments add to the richness of the aural architecture: the aesthetic sense of the space. Acoustic objects, such as instruments, vocals, or loudspeakers, are active objects in that they produce and project sound into the acoustic space. Room geometries such as walls and ceilings are also considered acoustic objects by Blesser and Salter (2007) as they transform the sound that is present in the architectural space and can be detected by the auditory process. Since they do not produce their own sound, Blesser and Salter define these as *passive* sound objects: ". . . any nonuniformity in the spatial distribution of sound implies the presence of an acoustic object or geometric anomaly, which is equivalent to an object" (Blesser & Salter 2007, p.154). Spatial inconsistencies that may be considered acoustic flaws in studios or traditional performance spaces can be explored and are part of the aural architecture. The phenomenology of aural space is complex; cognitive processes are subject to the listeners' own personal experiences, memories and culture. Each visitor brings their own unique personal perspective informed by their life experience and memories: the space becomes a place. Yi-Fu Tuan concurs: "What begins as undifferentiated space becomes place as we get to know it better and endow it with value" (Tuan 1979, p.6). Investigating the affective reactions to aural architecture is therefore difficult, it depends on the listener's personal history or temperament. The ear detects the sound and this sound is processed and transformed by the individual's personal perspective; as Kapchan (2017, p.5) describes it "passed through the prism of the ear and transformed with new colour and meanings". This has an implication for auditory spatial awareness and it is not something within the composer's control. Indeed, the composer brings their own perspective within any given setting. Furthermore, the affect may not be consciously registered by the listener. However, by integrating aural architecture into the compositional process the listening experience has the *capacity* to produce overt or subliminal affects: "Overt affect corresponds to strong feelings, emotions, whereas subliminal affect corresponds to subtle arousal, moods." (Blesser & Salter 2007, p.13).

The following section discusses some of my sound works and how I have developed a methodology to integrate site into composition.

3. Practice

Found Sounds *(2016)*

My practice began during a ten-day residency in a large former church called Left Bank Leeds, which was followed by an exhibition. The neo-gothic style of this site has many aural embellishments including an impressive, vaulted ceiling (see figure 11.1).

I spent some time (mostly at night) listening to the empty space and drawing inspiration from the audible features of the site as phenomenological properties (Spinks, 2015). I began to get a sense of what sounds would be effective to 'interrogate' the site and how they can be used in the compositional process. The material for *Found Sounds* was captured using a variety of microphones (including contact) and included found sounds, which were produced by activating materials by physically interacting with the site, as well as unheard or unnoticed sounds like creaks and bumps. I also used acoustic and electronic sources recorded in different locations. In this way the space processed the material, or 're-amped' it. Acoustic sources such as a small four-piece choir, percussion, and feedback guitar were also recorded in the site, and I took an empirical approach to positioning each source to locate the most suitable response from the building. The high, vaulted ceiling was particularly effective, interacting with the choir, and having found resonances, the singers improvised in a call-and-response with the building where notes were spectrally and temporally transformed. The electric guitar was also used to explore resonance. By using high gain settings, the orientation of the instrument emphasised different frequencies

FIGURE 11.1   Left Bank Leeds.

through feedback in one of the smaller side areas of the site. I used electronic sounds created by using a modular synthesiser to activate less accessible areas of the site. This proved to be an ideal tool for creating and manipulating sounds in a real-time, reflexive manner as parameters such as pitch, amplitude, and harmonic content can be easily and intuitively changed. Sounds were projected through loudspeakers, which were themselves orientated at aural embellishments and geometries that would transform the material tonally. Similarly, reverberation was used to develop the material created by the synthesiser, and changes in the envelope and tone of the sound activated the space in complex and enriching ways. For the exhibition, the recorded sounds were replayed in the space. The form of sound reproduction was an irregular multichannel system that I installed. The quiet or unheard sounds were presented on a large scale within the site along with the other sonic material. The composition is a collection of organised sounds, a collage of the recordings. These were presented during the exhibition via eight loudspeakers. Speakers were positioned in a quadraphonic arrangement at ground level using two public address (PA) systems. This allowed the sound to immerse the listeners. The PA system included a subwoofer under each mid-high speaker, with the latter being directed towards the walls and archways rather than facing the audience directly. It was important that the speakers themselves were not perceivable as the source and that the building itself seemed to be generating the sound. Additional speakers were placed upwards above the altar to extend the system to allow height in the presentation of the composition. As with the four floor-standing speakers, these were placed so the direct sound was not aimed at the audience. In this case they were aimed at the roof space so the sound seemed to be coming from above. The opportunities presented by this irregular site meant that although a stereo image would not be possible, a more interesting arrangement could be developed including height that would not be possible in many performance venues (Stefani & Lauke, 2010). The spatial mix of the piece was performed during the opening night using eight auxiliary sends on an analogue desk (illustrated in figure 11.2), which limited the number of channels available to me and also made control of the spatial positioning difficult. However, the work was well received and I made a multichannel recording of the spatial mix, which played during the continuing exhibition.

Listen, Hear! *(2018)*

Listen, Hear! (2018) is a ten-channel piece for a repurposed telephone isolation booth (as found in bars and nightclubs before the proliferation of mobile telephones). This work was selected from an open call for the Left Bank Leeds Summer Show in July 2018.

The physical space in this work is the telephone booth (see figure 11.3), and I wanted to emphasise 'spatial dissonance' as a way to engage the listener and create an imagined space (Smalley 1991). The shape and design of telephone booths is intended to reduce the spatial arena and isolate the occupant both acoustically and physically from noisy environments such as clubs, pubs, or railway stations. By installing multiple loudspeakers within the booth, I wanted to transport the listener to an imagined space, an imagined space within a physical space. Wishart describes "virtual acoustic space" as a means to project "an image of any real or existing acoustic space" (Wishart & Emmerson 1996, p.136). However, the virtual acoustic space need not be a recreation of an actual (realistic) acoustic space and can be any kind of imagined 'landscape'. As Blesser and Salter (2007, p.5) point out, the term aural architecture may also include the "creation of spatial experiences where a physical space does not actually exist". As the space I wanted to create is imagined it seemed appropriate to create material from sounds that are not readily experienced. Sounds can be inaudible (or imperceptible) due to differing conditions, for example above or below the range of hearing in terms of frequency or below the sensitivity of hearing. Sound may also be inaudible due

FIGURE 11.2 The spatial mix was created live during the exhibition.

FIGURE 11.3 *Listen, Hear!* (2018).

to inaccessibility, such as large or small spatial scales. Even when sound is audible, the way it is perceived depends on the attention given to it; some sounds are inaudible simply because there is no attention paid to them. To make sounds audible a personal listening space is created within the telephone booth to remove the visitor from the larger acoustic environment.

I chose to use sounds produced by the Hammond organ, an instrument that is central to my practice as a performing musician. I captured small acoustic sounds, such as the whirring of

FIGURE 11.4   Eight repurposed speakers are fixed within the booth.

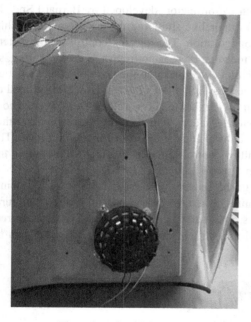

FIGURE 11.5   The sub and mid frequency transducers on the rear of the booth.

motors and the old valves warming up; contact microphone recordings of mechanical vibrations; and electromagnetic signals of the rich tonalities created by the tone wheels (Aldridge 1996). These sounds were then combined to form the final piece. The design of the booth made installing speakers difficult. It has a smooth, rounded surface within and I wanted to keep the outside of the booth untouched so it retained its original appearance. After some experimentation I decided to use drivers repurposed from Bose Sound Dock speakers. I attached these to the inside of the booth using Velcro as this is flexible enough to follow the contours of the internal structure of the booth and is actually very strong. The placement of the speakers is shown in the figure 11.4.

To increase the frequency range I added two transducers to the rear of the booth (shown in figure 11.5). These are not typical loudspeakers as they do not include a moving diaphragm, instead relying on direct contact with a surface, which then becomes the vibrating source. I found this to be very effective and included both a medium size transducer and much more powerful sub frequency transducer to enable much lower sounds and special effects to be produced in the booth.

Mixing audio for irregular arrays can be difficult, as panning is not always available in stock plugins for the number of channels or specific speaker positioning. Creating spatial gestures can be problematic without suitable hardware; I had realised from my first work *Found Sounds* that it can be challenging to manipulate the position of sound objects, particularly when height is added to the speaker system. I began to develop a suitable spatial control system using the graphic programming language Max (developed by Cycling '74). The software allows easy integration of hardware, and data can be filtered and scaled to suit a specific application. I decided the best approach would be to use MIDI data as this would be compatible with any digital audio workstation (DAW) and Max will run within Ableton as a 'Max for live' process. By using a hardware input, I wanted to be able to control the spatial position and amplitude of sound events and to be able to record the resulting data as MIDI so it could be recalled and edited if necessary. Kenton supply a 'DIY board' that allows analogue information to be encoded into digital MIDI (Kenton.com n.d.) For the analogue control I took some inspiration from the 'Azimuth Co-ordinator' developed by Bernard Speight for Pink Floyd in 1969 (Melchior *et al.* 2013). In their live show the device was used to shift sounds around the audience in a quadraphonic configuration. The joystick control in my system was originally designed for lighting applications, meaning there is no 'return to centre' spring, so it can be left in any position as required. The Kenton device picks up the position of the joystick via a simple X/Y output from the two potentiometers as in Speight's design. However, instead of directly controlling the analogue audio source, in my design the X/Y values are converted to a digital MIDI value 0-127. I also included two analogue faders, which could be assigned to other control parameters. In this case, I used one for signal level and the other for sub-speaker level. The MIDI data was scaled in a simple Max patch to allow control of the level and position within the irregular array. The finished controller is shown in figure 11.6 below.

For this work I wanted to produce an imagined space. As Blesser (2007, p.132) points out: "Whether modelling reality or creating a fantasy, the creator of a virtual space is an aural architect." A simulated spatial reality, he explains, is a 'surrogate' space – real spaces have real sound waves, but a virtual space can be created using signal processing. Virtual space can be

FIGURE 11.6 The joystick with volume and sub controls.

realistic in that it simulates a real space, and the listener will perceive a realistic space. The aural embellishments of real spaces are complex and to create a realistic simulated space would require tonal, temporal, and spatial processing to be effective. The virtual space could also be hyper-real and create an experience of space that extends the spatial dissonance by exaggerating the perceived size of the booth with the use of signal processing. In real spaces the reverberant field is not always completely diffused and omnidirectional, for example in irregular spaces or where there are adjoining spaces. Furthermore, in real spaces the aural embellishments add tonal richness to the perceived sound. In this work I needed to be able to simulate the same behaviour and also be able to extend this into more 'unreal' virtual spaces if required. The tonal, spatial, and temporal changes would be unique to each source in a real space and thus would have to be unique to each source in the booth – the space would transform the sound from individual speakers. This technique is described by Stefani and Lauke (2010, p.258) as "Spatial Mapping" – applying a spectral character to a spatial location by applying digital signal processing (DSP) to a specific channel output. It would not be realistic to simply add artificial reverberation to all the speakers, and furthermore, directivity would be compromised if the direct-to-reverberant ratio becomes too low (Everest and Pohlmann 2022). In my Max patch, I use convolution reverb plugins on each of the ten outputs, which allows me to tailor each speaker's response. I also added a function whereby the reverb can be spread to other speakers in proportion to their proximity to each other. This spread is adjustable, allowing for the reverb to be very localised or more diffused depending on the distance that is being simulated (Chowning 1977).

Tupperware Party *(2019)*

This was another commission for Left Bank, this time to accompany a different visual artist and a very different work: a large and luminous inflatable sculpture. The event flyer is shown in figure 11.7 below.

The sonic material was created using modular synthesis – the analogue controls are well suited to manipulating changes in dynamics, tone, and texture and have a gestural nature that lends itself to taking a spectromorphological approach to creating sounds. The initial work was a series of eight short works, each with a duration of ninety seconds followed by ninety seconds of silence. In this way the sound of the sculpture inflating became part of the work. The work was looped and ran for the whole month of the exhibition. For the closing night I reworked the shorter pieces into a longer, twelve-minute composition, with sixteen channels of audio. Speaker positions were chosen to get the most rewarding response from aural embellishments, particularly from the arches and vaulted ceiling. Once positioned, the sonic material could be presented through this site-specific array as a live diffusion on the closing night. This is not possible with conventional hardware. My work *Listen, Hear!* (2018) necessitated the development of a suitable system for multichannel mixing for irregular loudspeaker arrays and I wanted to include the same system to bring a performative element to this piece. As there are eight stems that make up this work there needed to be eight joystick controllers to allow 'players' to perform a live spatialisation. The software also needed to be developed to incorporate an effective means of controlling the height of each stem.

The overall concept is the same as described in the previous work *Listen, Hear!* but significantly developed to include two spatial controls, illustrated by two separate panels (see figure 11.8 above). The panel on the left shows the lower speakers in their approximate location within the site (channels 9 and 10 were actually at a mid-height above the church altar). The right panel shows the high speakers 11–15. The blue and red dot represents the position of the stereo sound source. The volume control is as before but the sub feed remote control that was used in *Listen, Hear!* was not

FIGURE 11.7   Inf23 by Michael Shaw featured my work *Tupperware Party* (2019).

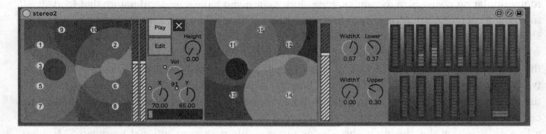

FIGURE 11.8   The developed Max patch includes height.

needed for this piece, so was reassigned as a crossfade between the lower and upper speaker sets (i.e. crossfading between the left and right panels). Sub frequency output was routed to a single sub speaker and its level controlled by the virtual fader to the right of the two panels on the software. The 'width' control changes the distance between the blue and red dots on both X and Y axes. This allows the stereo input signal to be represented in different positions, for example, across the speakers left-right or back-front. Alternatively, the width can be reduced to effectively collapse the stereo input to mono as required. In this work, the intention of the diffusion process is to place the sound materials at locations around the site, to allow the audience to experience the aural architecture through composition. The stereo image is not translated from the source material, meaning there is no 'sweet spot', and this is not the intention. The aural experience comes from the combination of the sonic material and its interaction with the site. Each performer was able to control their own stem via an individual joystick controller.

Music Boxes *(2019)*

I was asked by Left Bank to make a work to accompany *Critical Mass* (2019) by visual artist Eva Mileusnic. This is a multitude of small plaster casts that resemble pairs of feet and is a commentary on the transition of migrants across the globe (corridor8.co.uk, 2019). I decided to use music boxes as the medium for my sound work as the quiet, reflective sound is reminiscent of childhood. This was in response to some of the small cast feet, which seemed to belong to children. The artist Eva Mileusnic herself commented on the suggestion of footprints in the punched holes that produce the sound material as making a positive link to her visual work. The music boxes arrive supplied with blank cards allowing the user to punch their own melodic material. I looped the card so the music could play continuously and adapted the device by including electric motors to drive the mechanical boxes. It took some time to find motors that were quiet enough. I sourced motors from a till roll mechanism as I found these to be most suitable. I then incorporated passive infrared detectors (PIR), which are triggered by movement and activate the motors for a predetermined amount of time. This formed the instrument (see figure 11.9).

I made eight of these to be positioned around the site. The musical material is a combination of a simple melody, which is duplicated on four boxes. The other four have various gestural arpeggios as well as fragments of harmony that support the melodic material. In this way, the movement of visitors in the space defines and creates the sound work. This is an important part of the composition, as the movement of people around the exhibition creates infinite variations as each music box starts and stops at different times. As the last person leaves, the music boxes gradually stop.

In terms of the methodology, the planning stage had identified the sonic material early on and the concept of movement triggering the work was established from the beginning. However, the interaction of the sonic material and site was dependant on the placement of each of the eight music boxes. The placement had some practical limits. The visitors had to be able to interact with the

FIGURE 11.9   One of eight music boxes.

music boxes, meaning the visitors had to pass within the range of the motion sensors. By placing the music boxes in different locations the effectiveness of the interaction with site could be evaluated. Sonically, there were clearly identifiable locations where the higher frequencies were emphasised by reflections from the ceiling. The most effective were at the base of the pillars, which support smaller adjoining rooms that have an arched roof. These smaller areas are resonant in mid-high frequencies and the reverberation here is also perceptively different to the main area. The work consists of the quiet sounds of the music boxes alongside the sounds of visitors wandering around.

4. Conclusion

Through practice research I developed an empirical method to engage with site. The complexity of space and variations in loudspeaker response within that space necessitates experimentation with both the sonic material and the positioning of sources. My early work *Found Sound* (2016) highlighted the technical challenges of working with multiple electroacoustic sources in irregular configurations. The difficulties of manually operating the spatial position and level limited the compositional process, as it was not intuitive. This led to the development of software and hardware systems that were used for both *Listen, Hear!* (2018) and *Tupperware Party* (2019). This system demonstrates that there are innovative ways to interact with complex spatial acoustics, both real and imagined, in producing site-specific spatial works and, in the latter piece, performance practice. Repurposing and creating new tools is often part of practice research methodologies (Young 2016) and this was particularly apparent in my work *Music Boxes* (2019), where the movement of visitors to the site creates the piece. Empirical methods have shown the effectiveness of integrating site into the presentation of composed work. The methodology has developed to place aural architecture at the centre of compositional intent: ". . . sound defines the architecture and draws attention to it in a new way" (Philipsz quoted in Savitskaya, 2015). However, this raises an important question: how can these works be readily experienced outside the site for which they were intended? As Blesser and Salter (2007) point out, the experience of aural space includes memories and associations, or symbolic meaning derived from a particular location. Recordings cannot fully reproduce the experience of aural architecture. I have documented the works discussed in this paper using spatial recordings made with an ambisonic microphone (available at bobbirchmusic.com). However, site-specific work is entirely dependent on location, eloquently expressed by sculptor Richard Serra: "To remove the work is to destroy the work" (Kwon 2002 p.12). My methodology has developed substantially over the course of my practice and research, in part due to being able to experiment with different techniques within the same site – Left Bank Leeds. The cycle of presenting a work and then evaluating its success enabled me to identify issues in the workflow, for example multichannel diffusion in irregular arrays. Returning to the same site allowed me to develop technical solutions to this problem and even led to the development of a live diffusion performance practice that focuses on aural architecture.

It is clear that auditory spatial awareness (attentive, active listening) is essential to engage with the meaning, emotion, and symbolism contained within sound. It is not possible to force visitors to listen, but through presenting composition that actively engages with site, the aural architecture becomes integrated with the work. This allows the complexities of acoustic space and its potential for enhancement of musical material to be developed and by discussing these works I hope to inspire other practitioners.

References

Aldridge, H.B. (1996). Music's most glorious voice: The Hammond organ. *Journal of American Culture*, 19(3), 1–8.

Beranek, L.L. (1996). Concert Halls and Opera Houses: How They Sound. New York : Springer.

Blauert, J. (2008). *Spatial Hearing: The Psychophysics of Human Sound Localization*. Cambridge, MA: MIT Press.

Blesser, B. & Salter, R. (2007). *Spaces Speak, Are You Listening? Experiencing Aural Architecture*. Cambridge, MA: MIT Press.

Chowning, J.M. (1977) 'The simulation of moving sound sources', *Computer Music Journal*, 1(3), pp. 48–52. doi:10.2307/3679609.

corridor8.co.uk (2019). *Eva Mileusnic & Bob Birch: Critical Mass* [Online] Available from: https://corridor8.co.uk/article/critical-mass/ Accessed 1/5/22.

Emmerson, S. (1998). Aural landscape: Musical space. *Organised Sound*, 3(2), 135–140. doi:10.1017/S1355771898002064.

Everest, F.A. and Pohlmann, K.C. (2022) Master handbook of acoustics. New York: McGraw-Hill.

Gendreau, M. (2011). Concerted Structures. In *Site of Sound #2: of Architecture and the Ear*, ed. by Martinho, C. & LaBelle, B. Berlin: Errant Bodies Press, pp.33–42.

Kapchan, D. (2017). *Theorizing Sound Writing (Music/Culture)*. Wesleyan University Press. Kindle Edition.

Kenton (n.d.) Kenton Electronics DIY Boards [Online] Available at: https://kentonuk.com/product-category/diy-boards/ Accessed 1/2/17.

Kwon, M. (2002). *One Place after Another: Site-specific Art and Locational Identity*. Cambridge, MA: The MIT Press.

Lutz, J. (2007). Transpositions: Architecture as Instruments/Instrument as Architecture. In *Resonance: Essays on the Intersection of Music and Architecture*, ed. by Muecke, M.W. & Zach, M.S. Ames, IA: Culicidae Architectural Press, pp.169–190.

McCombe, C. (2001). *Imagining Space Through Sound*. In *Sound Practice Proceedings*. University of Edinburgh Sound Practice the 1st UKISK Conference on Sound Culture and Environments. Dartington: Hall Conference Centre and Dartington College of Arts, 2001.

Melchior, F. *et al.* (2013). On the use of a haptic feedback device for sound source control in spatial audio systems. In *Audio Engineering Society Convention 134*. Rome: Audio Engineering Society.

Sambolec, T. (2011). Recontextualizing Sound. In *Site of Sound #2: of Architecture and the Ear*, ed. by Martinho, C. & LaBelle, B. Berlin: Errant Bodies Press, pp.55–66.

Savitskaya, A. (2015). 'Sound has always been my primary tool – interview with Susan Philipsz about her sound installations and more', *ArtDependence* January 2015. Available at: www.artdependence.com/articles/sound-has-always-been-my-primary-tool-interview-with-susan-philipsz-about-her-sound-installations-and-more/ (Accessed 2/5/22).

Sheridan, T. and Van Lengen, K. (2003) 'Hearing architecture', *Journal of Architectural Education*, 57(2), pp. 37–44. doi:10.1162/104648803770558978.

Smalley, D. (1991). Spatial Experience in Electro-acoustic Music. In *L'espace du son II. Special issue of Lien*, ed. by Dhomont, F. (Ed.) Ohain: Musiques et Recherches. pp.123–126.

Spinks, T. (2015). *Associating Places: Strategies for Live, Site Specific, Sound Art Performance*. PhD thesis, University of the Arts London.

Stefani, E. & Lauke, K. (2010). Music, space and theatre: Site-specific approaches to multichannel spatialisation. *Organised Sound*, 15(3), pp.251–259. doi:10.1017/S1355771810000270.

Till, N. (2016). 'Sound Houses': Music, Architecture, and the Post-modern Sonic. In Yael Kaduri (Ed.), *The Oxford Handbook of Sound and Image in Western Art*. Oxford Handbooks, pp.163–191

Wishart, T. & Emmerson, S. (1996). *On Sonic Art*. Amsterdam: Harwood Academic.

Young, J. (2016). Imaginary Workscapes: Creative Practice and Research through Electroacoustic Composition. In *Artistic Practice as Research in Music: Theory, Criticism, Practice*, ed. by Doğantan, M. London: Routledge, pp.149–166.

12

"WE WENT FROM 'YES, YES YAWL', TAE 'WHO YOU TALKIN TAE?'"

Language and Authenticity in Scots Hip Hop

Sace Lockhart

1. Introduction

From its origin story in the Bronx, New York in the late 1970s and early 1980s hip hop has been classified into five cultural modes (pillars) of: MC'ing/rap (oral), DJ'ing (aural), breakdancing (physical), graffiti art (visual) and knowledge (mental). Each one of these modes has its own developmental history which can be analysed through existing theoretical frameworks. Rap music is also described as 'hip hop' along with the broader definition that incorporates the totality of the cultural modes.

As part of the Scottish hip hop community in the 1980s and '90s, I have personal experience of the development of hip hop culture in Scotland. In the early 1980s, I participated in the early breakdancing movement that swept the UK, culminating in the battles at the Plaza Ballroom in Glasgow, which brought together crews from across the country. The early 1990s saw the first vinyl releases from Scottish artists; my own group Two Tone Committee contributed in 1991 with a double-sided, 12-inch single released on 23rd Precinct Records, which was self-produced. As a DJ, I was involved in the first hip hop clubs in Scotland, the pinnacle of which was 'Blueprint' in Glasgow (1994–1996), which was the first club in the UK to have a custom-built skateboard ramp and the use of four turntables simultaneously. Using skills I learned through DJ'ing and recording/producing in the studio, I began experimenting with songwriting using melody and live instruments, combining this with the use of samples, which at the time of the confluence of analogue and digital technologies was difficult to execute. As part of the group NT, I was signed to major record and publishing deals that expanded my knowledge of the music industry. Although it was not altogether a positive experience for me I was keen to share my knowledge with younger artists. I have been an advocate for community music initiatives since the early 2000s, setting up two national mentoring projects that provide young people with opportunities that they would not normally have access to. The entirety of this experience forms the basis of my autoethnographic research which affords me the opportunity to access a network of key individuals that have helped shape hip hop culture in Scotland. In 2018, I produced and directed a documentary on Scottish hip hop for the BBC, presented by Orwell prizewinning author Darren 'Loki' McGarvey. The research, interviews and collection of artefacts formed the basis of my studies as well as the generation of a touring exhibition and the foundation of an archive.

DOI: 10.4324/9781003396550-12

The title of this paper refers to a quote from a Scottish hip hop artist named Davie Mulhearn AKA, 'Freestyle master'. This is a very distinct commentary on the global recognition of the foundation of hip hop and its transition to an authentic localised commentary. Mulhearn's comment encompasses the referencing of the early American hip hop phrase 'yes yes y'all, to the beat y'all' which was used by MCs to hype up the crowd for the DJ. The etymology of the phrase is debated with the possibility that DJ Kool Herc or Kid Creole from Grandmaster Flash and the Furious Five may have invented it, although it would seem to be ubiquitous at the time.

> Kid Creole was known for calling out, Yes yes y'all, and you don't stop, to the beat y'all and you don't stop.
>
> *(Katz, 2012:73)*

What Mulhearn demonstrates in this one line is the change historically, chronologically and ideologically in Scottish hip hop. It represents the move away from mimicking party rhymes to expressing localised content that reflected the times, and in the case of Mulhearn's native Glasgow. 'Who you talkin tae?' is a direct reference to a more confrontational and personal response that would be used as a remonstration to an offensive comment.

I explore the definitions of authenticity with language as a key indicator which I reference chronologically from the time of Robert Burns in the 1700s to hip hop culture in the 1980s and 1990s. I contrast character development in different time periods as an authentic device with diametric results relating to commercial success, that can be discussed in reference to class and cultural bias. Robert Burns as the 'Ploughman Poet' was seen as the saviour of Scottish language (Devine, 2012:294). I follow the influence of Burn's language which became highly valued to the 'slovenly' use of Urban Scots used by Scottish rappers. This examination is done within the hip hop tenet of 'keeping it real' that evidences the similarities of creative practices of Burns and Scottish hip hop artists. Creative licence is measured in acceptable processes across time relating to stature and in the case of hip hop's method of sampling, race. This feeds into the conclusion of authenticity referencing popular music and culture that explores the personal and socially agreed construct of authenticity as well as an additional definition of one that is 'living it out'.

2. Perversion of Dialect

In 1808 John Jamieson published the first Scots dictionary: "The Etymological Dictionary of the Scottish Language". The dictionary was split into two strands; the urban Scots of the Industrial Revolution, and rural Scots that became the romanticised language popularised by lovers of Robert Burns' works. On this, Simon Hall notes that there becomes a distinct difference in public opinion regarding the two forms of Scots:

> Many people began to take the view that rural Scots was good, while urban Scots was bad. Rural varieties of Scots began to be thought of as wholesome, traditional and authentic, while urban varieties were frowned upon, and thought of as 'slovenly perversions of dialect'.
>
> *(Hall, 2017:12)*

The phrase 'slovenly perversions of dialect' comes from a Scottish Education Department (SED) pamphlet published in 1952 entitled "English in Secondary Schools" which presented three categories of language:

1. An exemplar of English generally acceptable to educated Scots, (approvable)
2. Genuine dialect whether of the Borders or of Buchan (approvable)
3. Slovenly perversions of dialect (not approvable)

This third variety is described as 'bad Scots' and is mainly associated with the urbanised areas of Scotland, the areas that would predominate in Scottish hip hop language of the 1980s and 1990s.

> All of these observations about Scottish linguistic and cultural practices feed into the practice of rapping in Scotland. Perhaps the parallels of having a language and correct way of speaking imposed always results in the ludic subversion of what is 'correct'. (From an autoethnographic stance, my mother relays how when she was growing up she was repeatedly told that many Scots words or phrases were 'slovenly'. Thus, Scots becomes associated with negative connotations of being untidy and dirty.)
>
> *(Hook, 2018:50)*

As one of the best-known figures in Scottish history, the contribution of Robert Burns to the Scottish national identity was one of its saviour as referenced by Lord Rosebery at the centenary of Burns' death:

> The Scottish dialect as he put it, was in danger of perishing. Burns seemed at this juncture to start to his feet and reassert Scotland's claim to national existence; his Scottish notes range through the world, and he has thus preserved the Scottish language forever—for mankind will never allow to die that idiom in which his songs and poems are enshrined.
>
> *(Devine, 2012:294)*

Robert Burns used Scots language as an authentic device to develop his identity as the 'Ploughman Poet'. It is useful to consider how this version of Scots language became regarded as appropriate and highly valued (Broadhead, 2013:107). This 'authentic Scots' was then exported across the world as part of Scottish identity and, in literary terms, held up to be the standard.

> Nine-tenths of so-called Modern Scots is a concrete of vulgarised, imperfect English, in which are sparsely embedded more or less corrupted forms of the 'lovely words' with which Burns wove his 'verbal magic'.
>
> *(McNaught, 1901:27)*

It seems that any deviation from this romanticised version of culture was deemed unauthentic and low status (Hall, 2017:12). The corrupted forms of the 'lovely words' of Burns are the very definition of the Scots language on which Scottish hip hop was built as an authentic expression. In the same way that Burns' Scots had been exported across the world containing local references, a version of hip hop was commodified and marketed as a package at the intersection of visual and digital identities. The authentic image projected was also grounded in locality and had developed its own language. Artists following this movement had also to adhere to a certain code in order to be deemed authentic and any deviation from this can be read as unauthentic.

For many in Scotland, Robert Burns' poems were less about the aesthetic he created and more about the national identity he helped forge. Burns broke boundaries in what poetry and prose was meant to sound like in the United Kingdom and abroad, using Scots where standard English would have been favoured by his contemporaries. The authentic identity that MCs using a Scottish accent seek is identical to Burns. They in fact are closer to the intention of Burns than the facsimile poets that regurgitate romantic language because it is seen as authentic. MCs using their own accent are pulling on their local vocabulary in modern times in the same way that Burns did in the 18th century, only the vehicles for delivery are different. Whereas Burns used Scots in a vehicle dominated by standard English, Scots MCs use their own dialect in a vehicle dominated by (African) American English. The stark difference is the level of success each accent has at different points in time and a disparity in commercial recognition and success. Another key difference is the relationship between outputs and inputs. Burns' Scots was exported into America from Scotland and hip hop was imported into Scotland from America. The cultural nuances in these transactions are worth considering. The expression of human emotion transmitted through words and music at different points in time shares similarities in the causes and conditions that led to their inception. Both Burns and contemporary Scottish-language hip hop artists faced stereotypes, a form of hegemony and an asymmetrical relationship with English influence, at the same time as having bonds with contemporaries south of the border. Ironically, Scottish hip hop artists had to react to the kitsch, kaliyard romanticised version of Scotland that Burns helped to create. Hugh MacDiarmid wrote about this somewhat acrimoniously describing the period after Burns' death as:

> An apparently bottomless abyss of doggerel, moralistic rubbish, mawkish sentimentality and witless jocosity.
>
> *(MacDiarmid, 1973)*

Not only are Scottish rappers fighting against charges of appropriation but they also have to contend with the added expectation that Scottish people should have a certain cultural display as approved by historical gatekeepers.

The value in the expression of Burns and that of Scots hip hop artists has many different aspects relating to perceived success. Exploring the cultural output of both Burns and contemporary Scottish hip hop artists, the capsule that contains the outputs has different layers of influence but similar core devices in creating content. "Auld Lang Syne" was created in much the same way as hip hop music was, i.e. it was appropriated from earlier sources:

> The poetry and the music of the song as now known, have been developed from poetry and music which existed previously.
>
> *(Dick & Inglis, 1892:379)*

George Bannatyne in 1568 inserted an anonymous poem into his manuscript of poetry:

> The title of the poem "Auld kindnes Foryett," is in modern Scottish "[Should] auld acquaintance [be] forgot,"—the first line of all the subsequent poems on the subject.
>
> *(Dick & Inglis, 1892:380)*

Arguably Scotland's most famous song was made up from pieces of other parts of prose, appropriated. Hip hop's aesthetic of sampling and the bricolage created from DJ techniques evolved from taking snippets of other compositions to create new work. There are yet more similarities between Burns' appropriation of existing material and modern hip hop techniques. One glaring omission is the enforcement of copyright law. Without going into too much detail about intellectual property and common law in Scotland in the 18th century, it is fair to say that Robert Burns was never sued for 'borrowing' from earlier works. What is different here is the perception and moral judgement from its detractors about what sampling is and the tradition in folklore that's appropriate to borrow from, or use others' work as a basis for your own, as in the case of Robert Burns. There is of course evidence of a racial element:

> Overlying the debate over digital sampling is an ever-evolving discourse on the relationship between technology, race, and copyright. As African-American cultural production has historically influenced, and been influenced by, technological developments, so has the copyright system both adapted (and failed to adapt) to such advances.
>
> *(Cox, 2012:144)*

We can draw comparisons between both Burns and contemporary Scottish hip hop artist's creative processes. Although an African American culture, artists in Scotland took a great deal of influence from English artists, as did Burns in literary terms. Burns' image as the "God fearing ploughman" was certainly one he didn't play down but would seem to be more the development of a character than based in reality:

> Burns 'naive' persona was accurately disputed by John Logan as early as February 1787, in an unsigned article in the English Review, which insisted that 'Robert Burns, though he has been represented as an ordinary ploughman, was a farmer, or what they call a tenant in Scotland, and rented land which he cultivated with his own hands. He is better acquainted with the English poets than most English authors that have come under our review'.
>
> *(Leask, 2010:6)*

How different is this from the embellishment of character in contemporary hip hop? The idea that Burns used Scots language not only as a preferred meter but as a way of developing an authentic character resonated with the times, especially after the American Civil War. The American connection is again another thing that bonds both parties, one export, one import, one Scottish expression.

3. Authentic Hip-Hop

I look to some broader discussions of authenticity in popular culture and music, exploring the validation of the term that is described in the *Stanford Encyclopedia of Philosophy* as "Either in the strong sense of being of undisputed origin or authorship, or in a weaker sense of being faithful to an original or a reliable, accurate representation" (Varga & Guignon, 2023:1). One commonality in the spectrum of opinion on authenticity is what is considered 'real' is valued and what is considered 'false' is not. What is 'real' and authentic is a subjective discussion encompassing many academic disciplines and may be understood in the personal sense of being 'true to oneself' and a wider set of rules in a collective social identity.

However, as authenticity has been described as "ascribed not inscribed" (Moore, 2002:210), there exists a dichotomy between a cultural and commercial authenticity and a practitioner value-based one. Moore states his case for authenticity as not inherent in a performance, event, object or person but a socially agreed upon construct. He starts from the assumption that:

> Authenticity does not inhere in any combination of musical sounds. 'Authenticity' is a matter of interpretation which is made and fought for from within a cultural and, thus, historicised position.
>
> *(Moore, 2002:210)*

This is in opposition to the essentialised view of authenticity as a real experience where people speak the 'truth' of their situation. An important factor in establishing authenticity is the emphasis on the source of each 'genuine' experience. Using cooking as a metaphor, an authentic dish contains locally sourced ingredients, the recipe can be shared and individuals can add their own twist but the original dish is considered authentic which is fixed geographically in place.

The combination of sounds in music, I would argue, is important in the classification of being authentic. If we think of the Boom Bap sound of hip hop in the nineties or the amplification of a Les Paul guitar solo in the sixties, both would point towards a recognition and qualification of an authentic sound. Somewhere in between these two definitions of the personal and socially agreed construct lies a proposition that authenticity exists through cultural activity. Speers (2014:19) describes this as "living it out". In the case of Scottish hip hop, the 'authentic' development from the early 1990s was built upon the knowledge and understanding of African American pioneers but with a personal and ethical consideration. This authentic version of hip hop was true to the practitioners but could seem false to consumers or anyone outside the group.

> For one to be able to make a claim of authenticity, one has to know the culture from which hip-hop comes. Thus, by identifying the old school and back in the day as a period when a pure hip-hop culture existed, hip-hop community members invoke an authentic past that stabilizes the present.
>
> *(McLeod, 1999:144)*

The knowledge and respect from Scottish hip hop practitioners to the originators of the art form and this acting as a mark of authenticity developed into a personal examination of the notion of being truthful to oneself within a larger set of 'universal' rules qualified in the hip hop mantra 'keepin it real'.

Exploring the 'real' from a hip hop perspective, Morgan (2005) observes:

> The hip-hop mantra 'keepin' it real' represents the quest for the coalescence and interface of ever-shifting art, politics, representation, performance and individual accountability that reflects all aspects of youth experience.
>
> *(Morgan, 2005:315)*

The hip hop mantra 'keepin it real' that Morgan is describing grew in the 1990s as a reaction to the commercialisation of hip hop through moving image, music and fashion marketing, and a commentary on the removal of a culture from the community that created it. The disparate elements of hip hop (rap, DJing, breakdancing and graffiti) that had coalesced to become one movement, initially represented the localised opinions of practitioners. 'Keepin it real' became a reaction to what was deemed not authentic or 'fake'. This could be socioeconomical or ideological commentary and a changing interface between what had been 'traditional' hip hop and modernity.

Morgan's comment is a summary of a series of essays exploring hip hop in relation to Western philosophical tradition. She uses the metaphor of a rap battle to bring to the forefront some of hip hop's core tropes "rooted in its own classic battles of philosophy" (2005:205) constantly repeated and reframed as: represent, recognise, and 'come correct', i.e. to behave appropriately; to do something correctly. These three edicts represent the trinity of authentication that rather than being fixed are in a state of 'flow', navigating chaos to create balance. The process of establishing authenticity in the context of a rap battle can be compared to a dialectic exchange between two protagonists to establish 'the truth'. The 'real' for Morgan is framed within a set of core characteristics that are "sensitive to the complexities of place, time, and generation" (p.210) that would seem universal in the development of hip hop culture across the world. These characteristics include an understanding of:

- The foundations of hip hop historically.
- The common language of hip hop
- The local and personal history of hip hop
- The values, attitudes and beliefs of hip hop practitioners
- The representation of these beliefs
- The evaluation of ability and skill within these beliefs and consensus of opinion based round this knowledge.

Pennycook (2007) focuses on the localisation of hip hop exploring what could be contradictory forces in the assimilation of different cultures:

> One of the most fascinating elements of the global/local relations in hip-hop, then, is what we might call the global spread of authenticity. Here is a perfect example of a tension between on the one hand the spread of a cultural dictate to adhere to certain principles of what it means to be authentic, and on the other, a process of localization that makes such an expression of staying true to oneself dependent on local contexts, languages, cultures, and understandings of the real.
>
> *(Pennycook, 2007:103)*

The 'real' that Pennycook is discussing here incorporates Morgan's core characteristics with an added emphasis on localised language and accent. This is a commentary on the duality of cultures working somewhat in opposition to each other but adhering to hip hop's core principle of 'keeping it real'. From its inception, hip hop has focused on locality, borne out in culturally significant events such as the rap battles of 1980s groups Boogie Down Productions and the Juice Crew, where conflict arises over the authenticity of hip-hop artists within the different boroughs of New York City. The establishment of New York and, locally, The Bronx as the birthplace of hip hop had to go through the process of "represent, recognise, and come correct". This process of authentication is then repeated in endless iterations, from country to city, to communities across the world.

One marker of authenticity in hip hop, that has been investigated by academics is the use of localised language (Bennett, 1999, Pennycook, 2007 and Alim, 2015). My research focuses on the early 1990s when the seed of using a Scottish accent was planted in Scottish hip-hop culture, moving away from Americanisms and latterly in some cases, mimicry of English accents. Regarding the reasons for using your own accent, David Burnett AKA Big Div a rapper and producer from Paisley just outside Glasgow notes:

> I think it's important to use your own accent to come across as authentic. Personally, when I was doing it, it was to stay true to myself and to be honest with myself. Because if I want people to believe in me, but how can they believe in somebody that sounds fake.
>
> *(Burnett, 2018)*

Burnett is describing a value-based decision where authenticity, integrity and honesty are the key actors. There is a commitment to 'being authentic' and an acceptance of responsibility required to make this a reality. Burnett is describing a transformation to become his 'authentic self': one in which you live in a way that you are 'real' not fake. Laura Speers in her investigation of authenticity in the London hip hop scene explores this reflection in terms of two separate categories; rapper authenticity versus hip hop authenticity.

> Rapper authenticity and hip hop authenticity, although closely entwined, refers to the tension between what can be understood as the self versus the community, in that rappers want to assert individuality *and* demonstrate belonging to a group. Hip hop authenticity then refers to how practitioners have to follow certain tropes, practices and rules based on the culture's history to gain acceptance by the collective, and yet, on the other hand, be highly individual and original, which can be understood as rapper authenticity.
>
> *(Speers, 2014:12)*

The commitment to this 'authentic' self can of course be transituational, an individual can be 'authentic' in a number of fixed points across space and time that may not relate to each other. This is especially true of youth cultures and reminds us again of Morgan's commentary on the quest for coalescence and interface in the search to be authentic.

4. Regional Authenticity

As a recording artist and producer signed to major labels in the 1990s, I was always struck by the term 'regional' in relation to sales territories outside of London. Scotland as a country didn't sit well for me as a 'regional' description and this term permeated into descriptions of hip hop outside of the metropolis. This being said, Scottish hip hop in the 1990s fit in with the description of UK hip hop, at least from listeners outside of the UK's point of view. The distillation of hip hop culture from the global to the local manifested in the use of local language. MCs had to navigate the cultural origins of hip hop along with their own to produce a new category of authentication, one in which 'regional', local characteristics and commentary were valued. Chang (2007) believes hip hop needs to be understood as a culture in an anthropological sense. As such, the adoption of hip hop in Scotland as an African American culture is a perfect place to examine the term 'authenticity'.

The transition to an 'authentic' version of Scottish rap, can be described as one that moved away from simulating American culture and accent to using local culture and dialect. It was far from smooth in its development. Regarding this localisation process, Andy Bennett writes:

> The localisation of hip hop, rather than being a smooth and consensual transition, is fraught with tensions and contradictions as young people attempt to reconcile issues of musical and stylistic authenticity with those of locality, identity and everyday life.
>
> *(Bennett, 1999:180)*

Reflecting autoethnographically, I was part of the swell of hip hop that developed in the UK from the early 1980s. As a culture that incorporates music, fashion, art and dance there were usually very conformist ideas initially about what being authentic meant. This idea of authenticity relied on a visual or auditory anchor, for example the first wave of graffiti art was very much based on the book by Martha Cooper and Henry Chalfont, "Subway Art" (1984), which contained mainly New York styles. The 'authentic' graffiti styles of the mid to late 1980s in the UK replicated those photographs, with "Spraycan Art" (1987) offering a more global perspective a few years later.

The illegal practice of graffiti evolved as a counter-narrative to the construct of the urban environment (Evans 2014) and incorporated an element of mimicry from the styles that were presented to the young would-be graffiti practitioners (known as graffiti writers or simply writers) in the UK through two key publications: Subway Art and Spray Can Art.

(Evans, 2020:1)

Regional variations of graffiti in Scotland began from the same root text but developed differently according to the groups of practitioners and generally around one crew or individual that dictated the style. The Fallen Angelz graffiti crew from Edinburgh were prolific in the 1990s with a development of colourist graphic designs, Derm, one of the principal writers describes the influence of 'Subway Art' and 'Spraycan Art':

> There were two books which showed you what was going on, they were the bible and the new testament as they could be called but the whole idea of you looking at stuff and then you develop your style off of that. I suppose isolations a good thing and you worked off other people's ideas and played – evolved collectively which I think is why in latter years when Poise One, Mor and myself were painting walls it was much more, let's play have fun and do something different you know?
>
> *(Scott, 2018)*

The localised style expressions varied in cities across Scotland as demonstrated in Glasgow, led by Mak One who concentrated on 3D lettering with graphic character embellishment. Mak began writing his name as young as 10 years old as part of the ubiquitous gang culture that went back to the 1940s.

> There has been gang Graff in Glasgow since I think the 1940's or something like real stylised like gang mentions, you never had them anywhere else in the U.K and I've spoken to some guys from East L.A and sent them pictures of old Glasgow stuff and they are like that's amazing.
>
> *(Kennan, 2018)*

The tradition of gangs marking local territory was carried on by Mak who found a symbiosis in the explosion of graffiti art in the 1980s. Gang graffiti was a precursor to the term 'tagging' that became part of the evolution of graffiti writing, in the west of Scotland the form of marking your name (or gangs name) was known as writing a 'Menshy' an abbreviation of 'mention'. The transition to writing a more stylised form of graffiti art seemed an obvious progression for Mak along with the lack of facilities and opportunities in the Govanhill area of Glasgow in the 1980s.

> I was pretty driven with Graff, there was nothing much else in my life back then you know what I mean it was pretty grim times wunnit, we might have been doing what normal kids were doing and going into the city centre and going to pubs and clubs and stuff but we were hanging about, building a fire and drinking wine and painting, there wasn't an awful lot else for us to do.
>
> *(Keenan, 2018)*

Mak was known as 'King of the line', an acclaimed title of notoriety that was earned through his proliferation of work and judged by his peers. Other terms such as 'Style master', which was used in much the same way as traditional art movements, meant that key individuals were responsible for advancing certain styles. Referring to the same foundational sources Mak cites early New York writers as his inspiration:

My stuffs just coming from New York basically, Dondi, Skeme and SEEN aw them guys I was just trying to emulate them you know what I mean, trying to construct letters how they done it, back then we were all just trying to copy somebody we thought was dead good, put your own wee twist on it.

(Keenan, 2018)

As opposed to Edinburgh, Glasgow's localised style was led by Mak One who was in turn influenced by what could be described as the localised New York graffiti scene. There is a continuation of the development of form from local to global that is replicated across the 'pillars' or modes of hip hop culture.

In the 1980s, Penilee, a housing scheme in southwest Glasgow, was where arguably Scotland's premier breakdancing crew The Glasgow City Breakers were resident. The crew went through a similar trajectory in the recognition of 'authentic' moves. The initial wave of breakdancing was copied from early moving images in theatrical releases, music videos and in some cases adverts. The fashion aspect of breakdancing incorporated the authentic look of American hip hop, that of suede Pumas and fat laces. Interestingly the availability (or lack) of the 'authentic' items in hip hop – vinyl, paint and fashion – led to creative licence and interpretation. In the case of breakdance fashion, for example, the authentic items were not affordable to working-class youth, so they had their own take on what was authentic. This was demonstrated by the Glasgow City Breakers' endorsement of Kappa (sportswear) tracksuits, for example, and the use of sewing fabric for 'Fat laces'.

Hip hop music in Scotland developed based on the knowledge of how it was created in America. This followed traditionally from two turntables 'spinning back' a section of a record called a break to create a breakbeat, to the first samplers and drum machines. At its inception, Scottish hip hop was considered authentic if it managed to sound anything like its American counterparts, this included for many the replication of an American accent. Scottish hip hop can be described as an acculturation of an Americanised cultural practice with foundations in African American culture. It is a hybrid culture that holds true to its foundational narrative of African American identity while expressing local experience. Scottish hip-hop artist and academic, Dr Dave Hook, whose group Stanley Odd was at the forefront of a new era of Scottish hip hop bands that used their own accent to a backbeat of real instruments and samples, notes:

I think as a writer, or as an artist or a creator, the longer you spend within whatever your medium is. The more you feel you need to understand its history and its social and cultural evolution.

(Hook, 2018)

The first Scottish hip hop vinyl releases in 1990/1991 began the transition to using a Scottish accent. Two of the releases by Sugar Bullet (1990) and Dope Inc. (1991) were protest songs against the Community Charge or Poll Tax which was a system of tax introduced by Margaret Thatcher's government in Scotland in 1989 a year before England and Wales. The songs by Sugar Bullet and Dope Inc. were rebellious in content, with what could be described as pro-Scottish sentiment. It was clear that Scottish hip hop from its inception was politically aware and the commentary was part of a larger authentic identity.

The Celtic Diaspora could be a useful frame for political strategy, a "strategic essentialism" (Spivak 1996, 214) used to perform strong identities in nations like Wales, Brittany, Ireland, and Scotland. Often, this family of otherness in hip-hop is defined ethnically—difference becomes a strong defining factor in the rhetoric of hip-hop, as the style was originally conceived as African American music which could be compared to the subaltern states of other ethnic minorities in a given country (e.g., Turkish German, French African, or aboriginal Australian).

(Williams, 2020:94)

Williams is drawing a comparison between a 'Celtic Diaspora' and other ethnic minorities in their own expressions through assimilation with hip hop. The exploration of identity through the engagement of hip hop culture resulted in a classification of authenticity that was based round local cultural knowledge. In contrast to the codification of authenticity in American hip hop, through reaction to mainstream success (Blair, 1993:499), Scottish hip hop had no comparable financial stimulus. Authenticity from a Scottish perspective began firstly with a knowledge of the American pioneers and the ability to reference these cultural and artistic outputs. In an academic sense, it was an oral and stylistic referencing of practice that was memorised. This then fuses over time with the rationalising and grafting of the local onto these existing markers of authenticity.

5. Conclusion

In 2018, I interviewed Scottish rapper Davie Mulhearn AKA 'Freestyle Master' for my documentary on Scottish hip hop for the BBC. The interview was comprehensive in Davie's recollection of hip hop history and his contributions to the culture. It was only when I was exploring definitions of authenticity and how this related to identities in Scottish hip hop that his phrase and title of this paper really resonated. As much as I have explored the theme of authenticity academically I am struck by the simplicity of Davie's sentence that summarises the assimilation of an African American cultural form to integrate with Scottish identity and culture. I became interested in our values towards what is acceptably 'authentic' and how this changes over time as well as our understanding of what being 'authentic' actually is. Being immersed in hip hop culture since the early 1980s I see the value in Davie's comment but this would be completely lost on someone that doesn't understand the context. I contrast this with the perception of Scotland's famous poet Robert Burns, whose use of Scots language required a glossary for non-speakers, yet his development of an authentic character was highly valued.

I am interested in the use of language as a gauge of authenticity; this I believe is a universal recognition of an individual sense of place and one that is rooted in locality. I explore Robert Burns as the 'Ploughman Poet' and his use of Scots language to project an authentic image. I evidence a direct line between the 'lovely words' of Burns and the 'slovenly perversion' of Scottish rap language, noting the irony in both parties' desire to be authentic with starkly different public responses and commercial success. I am fascinated by the seemingly simultaneous movement of Scottish (and UK) rap artists' shift to use their own accent within the hip hop tenet of 'keeping it real' which has developed a key set of 'universal' rules summarised by Morgan (2005).

My research investigates the use of language as a marker of authenticity which in hip hop is a global trait that is manifested at a local level. (Pennycook, 2007). I explore the transition in Scotland with MCs using their own accent, non-standard English and Scots language auto ethnographically through observations and interviews. In my analysis, authenticity in Scottish hip hop is not fixed but evolves; we see this in the initial replication of authentic American influence through the ethical consideration of 'keeping it real' by using local cultural influences, language and accent. Authenticity for me then is both personal and a socially agreed construct that is fixed and subject to change; the definition is personal for the subject and subjective for the analyst, something that is 'living it out'. I return to the title of my paper which summarises in one sentence the entirety of my research with a sense of Davie's intrinsic confidence, 'Who you talkin tae?' indeed.

References

Alim, H.S. 2015. 'Hip hop nation language: Localization and globalization'. *The Oxford Handbook of African American Language*, 850, pp.850–862.

Bennett, A. 1999. Rappin'on the Tyne: White hip hop culture in Northeast England–an ethnographic study. *The Sociological Review*, 47(1), pp.1–24.

Blair, M.E. 1993. Commercialization of the Rap Music Youth Subculture. *The Journal of Popular Culture*, 27, pp.21–33.

Broadhead, A. 2013. The Language of Robert Burns: Style, Ideology, and Identity. Bucknell University Press.

Burnett, D. 2018. 'Loki's history of Scottish hip hop'. Interview by Sace Lockhart [in person], 5th September.

Chang, J. 2007. Can't stop won't stop: A history of the hip-hop generation. St. Martin's Press.

Cox. 2012. Not in Court cause I Stole a Beat : the Digital Music Sampling Debate's Discourse on Race and Culture, and the Need for Test Case Litigation, University of Illinois Journal of Law, Technology and Policy 141 (Spring 2012).

de Paor-Evans, A. 2020. Urban Myths and Rural Legends: An Alternate Take on the Regionalism of Hip Hop. *Popular Music and Society*, 43(4), pp.414–425.

Devine, T.M. 2012. *The Scottish nation: A modern history.* Penguin UK.

Dick, J. and Inglis, A. 1892 November. 'Auld Lang Syne'-its Origin, Poetry, and Music. In *Proceedings of the Society of Antiquaries of Scotland* (Vol. 26, pp. 379–397).

Hall, S. 2017. *History of Scots*: https://education.gov.scot/nih/Documents/HistoryOfScotsMar17.pdf

Hook, D. 2018. *An autoethnography of Scottish hip-hop: identity, locality, outsiderdom and social commentary* Doctoral dissertation, Edinburgh Napier University. Available at: www.napier.ac.uk/research-and-innovation/research-search/outputs/an-autoethnography-of-scottish-hip-hop-identity-locality-outsiderdom-and-social (Accessed:10 June 2018).

Katz, M. 2012. *Groove music: The art and culture of the hip-hop DJ.* Oxford University Press on Demand.

Keenan, T. (2018) 'Loki's history of Scottish hip hop'. Interview by Sace Lockhart [in person], 27th September.

Leask, N. 2010. *Robert Burns and pastoral: Poetry and improvement in late eighteenth-century Scotland.* Oxford University Press.

MacDiarmid, H. 1973. *Introduction to Robert Henryson's Testament of Cresseid*, Penguin Classics, p.7.

McLeod, K. 1999. Authenticity within hip-hop and other cultures threatened with assimilation. *Journal of Communication*, 49(4), pp.134–150.

McNaught, D. 1901. The raucle tongue of Burns. *Annual Burns Chronicle* 10, pp.26–37.

Moore, A. 2002. Authenticity as authentication. *Popular Music*, 21(2), pp.209–223.

Morgan, M., Darby, D. and Shelby, T. 2005. *After . . . word! The philosophy of the hip-hop battle*, Open Court.

Pennycook, A. 2007. Language, localization, and the real: Hip-hop and the global spread of authenticity. *Journal of Language, Identity, and Education*, 6(2), pp.101–115.

Scott, D. (2018) 'Loki's history of Scottish hip hop'. Interview by Sace Lockhart [in person], 30th September.

Speers, L. 2014. *Keepin'it real: negotiating authenticity in the London hip hop scene* (Doctoral dissertation, King's College London (University of London)).

Varga, S. and Guignon, C. 2023. "Authenticity", *The Stanford Encyclopedia of Philosophy* (Summer 2023 Edition), Edward N. Zalta and Uri Nodelman (eds.), https://plato.stanford.edu/archives/sum2023/entries/authenticity/. (accessed: 12 June 2021).

Williams, J.A. 2020. *Brithop: The politics of UK rap in the new century.* Oxford University Press.

13

MIGHT AS WELL BE SWING

On the Use and Misuse of Quantisation in Hip-Hop Production

Zachary Diaz

1. Introduction/literature Review

In many instances, we talk about musical innovations using the term "revolution". Musical instrument or technology companies will market their equipment as "revolutions" in music creation, advertising the ease or satisfaction one may have in using their product. Major historical shifts in the creation, production, or influence of musical recordings tend to be talked about as being "revolutionary", such as the Beatles' use of double tracking or Public Enemy's creative use of multiple dissonant samples. Un-quantisation is one such example of a musical revolution and is described as such by scholars such as Anne Danielsen (2006) and Dan Charnas (2022). Charnas credits this explicitly to J Dilla in his book *Dilla Time*, with the thesis of the book itself focusing on the revolutionary influence J Dilla's music has had on contemporary performance and interpretation of rhythms.

Revolutions are usually only best understood in retrospect, however, as the influence of said revolution will manifest itself over the course of history. How we interpret revolutions may also change as history progresses, with new information revealed by scholars and thrown into discourse by communities as to the value of new historical knowledge. The development of quantisation in the early 1980s, for example, can be argued to be just as revolutionary as its opposite. Other producers besides J Dilla frequently experimented with un-quantising. How and why producers such as J Dilla choose to quantise may be beyond just the desire to "humanise" one's rhythms. Finally, over the course of its development and popularity over the past two decades, the act of un-quantising may be far from radical or revolutionary, and in some communities or musical subgenres, may be quite ordinary, typical, or even drab.

How both use and misuse is interpreted is highly dependent on previous examples of use or misuse of a musical idea, both historically and culturally. Cultural aspects of use or misuse may tie themselves to notions of authenticity, as well as the desires of either fitting in or standing out among a particular genre or local music scene. The concepts of correct uses of a musical tool or idea may evolve into traditions or customs and may do so quite quickly. As digital music technology has rapidly evolved over the course of the last four decades, certain customs and traditions have both been established, broken, and reestablished. Joseph Schloss's (2014) ethnography of hip-hop producers in *Making Beats* is a prime example of this, with all the hip-hop production "rules"

DOI: 10.4324/9781003396550-13

highlighted by Schloss both being established and critiqued by both fans and fellow producers in the span of only slightly more than a decade.

2. On the Practice of Quantisation

One of the most significant and unique characteristics in all Afrodiasporic music is the drum pattern. Because of this significance, one of the main goals for this chapter is to emphasise that how musicians collectively view specific approaches to drum pattern creation is shaped both by the specific techniques themselves, usually created or popularised by specific individual producers, and by how drum patterns have evolved or developed historically throughout a genre or genres of music. This understanding of both synchronic and diachronic histories of drum patterns are crucial to understanding both their origins, functions, and influence. One major player in this is the collective historiographies that make up how hip-hop culture interprets and understands the influence of drum patterns on hip-hop production, adding to what Nate Harrison (2004) calls the "collective audio unconscious". Within this collective unconscious, certain musical ideas become so ubiquitous to an audience through popularity and exposure that audience members may find it difficult to identify the idea or its related historical roots within a recording. In popular music studies discourse, it is common to attempt to transfer a musical idea from the collective unconscious to the conscious, making aware or bringing to light the historical and cultural background of a specific musical trope to wider audience. Just like any form of dissemination of information related to popular culture, however, biases and misunderstandings may occur. This can result in the misidentification of a musical idea, such as the wrong identification of a sample's source in a hip-hop beat, or a producer being uncredited (or wrongly credited) in a production they may have participated in. I can personally attest to this phenomenon, as I have had numerous experiences with students or colleagues misidentifying what is interpreted by them as facts of hip-hop history that they may have read from articles or videos online, in which said articles' sources may simply be from long-standing rumours from the industry or from various fanbases. Before discussing misconceptions of quantisation, however, it is important to discuss its origins and original dissemination among music technologies in the 1980s.

The term "quantisation" refers to the option of quantisation found among many digital samplers and audio workstations, in which the software can "quantise" or correct any rhythmic inaccuracies when recording or inputting rhythmic patterns into a MIDI grid. Throughout its early development in the 1980s, the quantisation tool became useful for electronic musicians and producers to record rhythmically accurate drum patterns or synthesiser melodies onto exact rhythmic divisions or subdivisions and was used frequently as a tool for early hip-hop producers such as Marley Marl and Pete Rock, using early digital samplers like the E-Mu SP-1200. Marley Marl's 1985 single "Marley Marl Scratch" and Pete Rock & CL Smooth's 1992 single "They Reminisce Over You (T.R.O.Y.)" are solid examples of this, with both tracks featuring fully quantised drum patterns in eighth and sixteenth notes. Figures 13.1 and 13.2 below show the drum patterns in "Marley Marl Scratch" and "They Reminisce Over You" respectively in both sheet music and a MIDI grid, highlighting the precise rhythmic groove in each pattern due to quantisation.

This quantisation tool has been present among digital sampling equipment such as the Akai Professional MPC and E-Mu SP-1200, and though they differed in their respective interface formats (both between each other and in DAWS such as Ableton Live), their approach to quantisation by the user remained the same. As J Dilla began to produce during his teenage years using equipment such as the MPC and SP-1200, he felt this use of quantisation to be restrictive and would often turn off the quantisation tool that was found on many of the samplers and digital workstations

FIGURE 13.1 Sheet music and MIDI grid of "Marley Marl Scratch" 1:25–1:35.

FIGURE 13.2 Sheet music and MIDI grid of "They Reminisce Over You" 0:35–0:44.

he used. This desire to create a more natural or "human" feel to his drum patterns became, as stated earlier, a signature aspect of his production. Although other artists moved away from using quantisation as mentioned previously, there were times where this was controversial among fellow producers and other musicians, with certain producers seeing un-quantisation as uneven or off-tempo. In one case, J Dilla's use of un-quantisation led to physical confrontation during his collaboration with Los Angeles hip-hop group The Pharcyde during the production of their 1995 album *Labcabincalifornia*, with Pharcyde member Slim Kid Tre, stating:

> Fat Lip and I fought physically over the way Jay Dee originally programmed 'Runnin'.' Fat Lip went in and reprogrammed every straight beat because Fat Lip was all about having the beats a certain way. I fought for it to be the way that it was because I was a stickler about people's creative input – that's what we hired him for.
>
> *(Houghton, 2006)*

This is a prime example of how any sort of perceived radical departure may result in reactionary responses, leading to criticism, cynicism, and possible outright rejection of the new development. This method in and of itself is Signifyin(g) because of the act of obscuring of meaning and intention, being both described as "drunk" or "sloppy" by listeners while also being precise and specific in terms of implementation by J Dilla. Figures 13.3 and 13.4 below show selected excerpts of instances of un-quantised drum patterns in the tracks "Drop" by The Pharcyde (1995) and "Get a Hold" by A Tribe Called Quest (1996), both in sheet music showing approximated rhythms and a MIDI grid providing more detailed visualisation of the un-quantised drum or drums and their locations within the subdivisions of each beat.

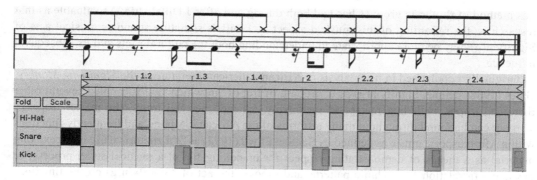

FIGURE 13.3   Sheet music of approximate rhythms and MIDI grid of "Drop" 0:42-0:47.

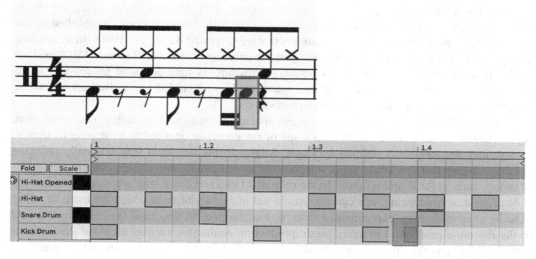

FIGURE 13.4   Sheet music of approximate rhythms and MIDI grid of "Get a Hold" 0:10-0:12.

3. Un-quantising as Signifyin(G)

In Simon Reynolds' (2009) article "The Cult of J Dilla", he lists the late producer's un-quantised rhythmic drum patterns as one of the three main characteristics of his signature sound. As stated throughout this chapter, it is difficult to understate the importance of this aspect of his production style, as it has led many musicians influenced by J Dilla's music to rethink the ways in which they approach rhythm in their own performances or compositions. This is a common sentiment among scholars and musicians who are fans of J Dilla, with jazz pianist Robert Glasper in an interview claiming that his un-quantised rhythms have influenced how both producers as well as performers of acoustic instruments have interpreted rhythmic patterns in live performance. In *Dilla Time*, Charnas (2022) also emphasises a similar viewpoint to Glasper's and centres the focus at which he views J Dilla's life and creative output through his use and experimentation with un-quantised rhythms, further emphasising the significance of this act. Though he does not frame it specifically through the lens of Signifyin(g), Charnas makes sure to highlight the sphere of influence that this method of rhythmic grooves has had both during and after J Dilla's life. As valuable as that research is, the reframing of this musical aspect of J Dilla's signature sound as musical acts of Signifyin(g) may be more appropriate in understanding both the cultural roots and its subsequent influence.

Originally popularised in his analysis of African American literature in the *Signifying Monkey*, Gates (1988) defines Signifyin(g) as a "troping of tropes" in which "are subsumed several other rhetorical tropes, including metaphor, metonymy, synecdoche, and irony (the master tropes)". This act can be found throughout many artforms of the African diaspora, and frequently functions within Afrodiasporic music as defined by Samuel Floyd (1997) as "the rhetorical use of preexisting material as a means of demonstrating respect for or poking fun at a musical style, process, or practice". In relation to rhythmic patterns and groove, the act of "Signifyin(g) on the timeline" originally coined by Floyd is further defined by Tim Hughes (2003) as "where 'timeline' refers to the underlying square, steady, and above all else European metric pulse". Through the process of creating un-quantised rhythms, producers such as J Dilla participated in Signifyin(g) on the timeline through the breaking of rhythmic expectations, with this expectation being the steady pulse of a backbeat found in much popular music, especially from the tropes established by the steady patterns of drum machines and samplers that were created during the 1980s. In developing his signature sound, J Dilla both Signified upon not only the timeline or rhythmic MIDI grid but implicitly upon other innovators of Afrodiasporic rhythm, as funk and soul groups and artists such as James Brown and Bootsy Collins implemented live-recorded improvisations and subtle "Signifyin(g) on the timeline" in their own productions.

As further described by Floyd, certain musical characteristics such as swing in jazz are more appropriate to be considered simply as a trope of the genre, as opposed to a "troping of tropes" as described and originally defined by Gates. Although this statement was originally referring to jazz, it seems that one could apply this to other more contemporary genres that use elements of swing, such as R&B and hip-hop. In the case of J Dilla and those who attempt to emulate his style of un-quantised swing, this can become an act of Signifyin(g), in which the trope of swing is being expanded upon, experimented with, or transformed. In this case, the "timeline" would be the rhythmic grid found in many music production equipment such as the Akai Professional MPC or DAW, and the Signifyin(g) would be the micro-rhythmic discrepancies and un-quantised drum patterns J Dilla implemented. In *Dilla Time*, Charnas emphasises this very intentional implementation, stating:

> [J] Dilla's rhythms were not accidents, they were intentions. Yet even the biggest fans of his style initially heard them as erratic. Why? Their reactions had everything to do with those rhythms defying their expectations. To understand the music of J Dilla, we must examine that process of subversion.
>
> *(Charnas, 2022, p. 45)*

Producer and drummer Questlove (Ahmir Thompson) further supports this claim by Charnas by describing his experience of hearing J Dilla's drum production on The Pharcyde's "Bullshit" for the first time in 1995 in his memoir *Mo' Meta Blues* (2015), stating that:

> . . . [J] Dilla was just going crazy on the kick [drum] pattern. At that moment, I had the same reaction I do to anything truly radical in hip-hop. I was paralyzed, uncertain how to feel. Usually, if I go over the top with my approval for an album or a band, it turns out to be a solid achievement . . . but like I said, if I'm brought up short by a piece of artwork, if I'm conflicted,

confounded, and made uncomfortable, nine times out of ten that thing will change the course of history. That's the feeling I got when I heard Dilla's kick pattern on "Bullshit."

(Thompson, 2015, p. 156)

In this instance, Questlove's understanding of typical kick drum pattern rhythms was challenged upon his first listening of "Bullshit". The un-quantised nature of the kick pattern also happens sporadically in the recording, keeping the listener guessing as to which kick drum hits will occur directly on downbeats or upbeats and which will occur somewhere in between. This experience from of Questlove shows him observing J Dilla's act of "Signifyin(g) on the timeline". Based on these examples above, how can one further interpret the wider historical evolution of this act of Signifyin(g), both by J Dilla and those producers who also participate in this act?

4. Historicising Un-quantisation

Terms like Danielsen's (2010) concept of "balanced imbalance" accurately describe the un-quantised rhythms found in much of J Dilla's discography (as well as the wider act of "Signifyin(g) on the timeline"), as the rhythmic spaces between each division and subdivision of the beat contains rhythmic discrepancies not only within each instrument but between each instrument being performed throughout the recording. How these rhythms have evolved historically and how J Dilla's influence fits within the wider evolution of un-quantisation is necessary to fully understanding its evolution. The diagram below (Table 13.1) shows this evolution separated into several eras of un-quantisation and its relation to the evolution of music technologies.

This table highlights how the practice of "Signifyin(g) on the timeline" through un-quantising developed historically. From about 1983 to 1987 there is the development of MIDI and the release of E-Mu SP-1200 and Akai MPC, where quantising in relation to the MIDI grid was developed by a collaboration of engineers Dave Smith and Chet Wood (Harkins, 2019). From 1988 to 1993 a period of experimentation during what is known as hip-hop's "golden age" occurs, with producers from Public Enemy's Hank Shocklee and Wu-Tang Clan's RZA attempting to avoid using quantisation and have loops of samples slightly un-synchronised to create a slightly uneven rhythmic groove. From about 1995 to 2005 popularisation of un-quantisation by J Dilla occurs, with even further experimentation through neo-soul artists such as Erykah Badu and D'Angelo. Finally, from 2006 to today is the post-Dilla era of hip-hop production, where instrumental hip-hop becomes a

TABLE 13.1 Historical timeline of un-quantisation.

1983–1987	Development of MIDI, Release of E-Mu SP-1200 and Akai MPC, where quantising in relation to the MIDI grid was developed.
1988–1994	Golden Age Experimentation from Bomb Squad and RZA. According to Jeff Chang (2005), Hank Shocklee of the Bomb Squad mentioned that he would avoid using quantisation, RZA would attempt to have loops of samples slightly off to create a slightly uneven groove.
1995–2005	J Dilla era, popularisation by J Dilla, further experimentation during Soulquarian era (artists such as D'Angelo and Erykah Badu), un-quantised grooves are heard more in R&B and Neo-Soul as well
2006-Present	"Post-Dilla" era of hip-hop production. Instrumental hip-hop becomes more of a growing subgenre through the LA beat scene, stuff like lo-fi hip-hop becoming increasingly popular and a standardisation of un-quantisation takes place.

growing subgenre through the Los Angeles beat scene and lo-fi hip-hop, further popularising and standardising un-quantised grooves.

In relating the phenomenon of quantisation to the framework of Hegelian dialectics as described by Maybee (2020), one could view quantising and un-quantising as a dialectic, with quantising acting as the thesis, un-quantising acting as antithesis, and a synthesis featuring the various spectrum of quantisation and un-quantisation decisions available to the user. Through this lens, then, one can see how significant J Dilla was to the development of programmed rhythms in popular music over the past few decades, acting as a leader of this antithesis and leading to further cultural and musical development. As mentioned earlier, books such as *Dilla Time* (Charnas, 2022) feature this as its central focus, portraying Dilla as a figure which reinvented rhythm which radically shifted understandings of groove in popular music.

This Hegelian viewpoint of J Dilla, however, may not give us the full picture when it comes to fully understanding how un-quantisation developed both culturally and historically. In a broader sense, it may be more appropriate to view J Dilla's contributions as a part of a wider radical tradition found within Afrodiasporic musical culture, Signifyin(g) both on the timeline of the beat and the historical timelines of musical history contextualised historically within what Cedric Robinson (2020) referred to as the "Black radical tradition". This leaning towards Western philosophical or cultural frameworks for understanding J Dilla's music is nothing new and can be found by journalists such as Rob Fitzpatrick (2011) or Jordan Ferguson (2014) calling J Dilla the "Mozart of hip-hop". Quantisation and its relationship to hip-hop production, however, may be understood better beyond what Philip Ewell (2020) calls the "white racial frame" of music theory and history. Viewing figures such as J Dilla as "great masters" similar to Western European composers may limit our understanding of how music creation and cultures such as that of contemporary hip-hop production accurately function. The cultural hegemony of Western historiographies has led many to interpret both un-quantisation and its relationship to J Dilla as tied to great innovations created by the more traditional viewpoint of "great men of history". As scholars and musicologists, however, we must acknowledge the limiting factors of this viewpoint. As shown in earlier examples such as Fitzpatrick or Ferguson, writers within the fields of music production and hip-hop (both academic and journalistic) have perpetuated this mythologisation.

In Charnas' (2023) article "Was Hip-Hop Really Invented 50 Years Ago?", he highlights that the use of DJ Kool Herc's first party in the south Bronx as a singular "birth of hip-hop" may be an historical oversimplification. The same is true for J Dilla and his use of un-quantisation, as no specific artist, song, or event can capture how or when a musical idea developed. Within this chapter, the use of Hegelian dialectics as a method of historical understanding and interpretation for the evolution of quantisation is implemented. This may be appropriate when trying to interpret the historical and technological development of quantisation, as the technology itself is closely linked to how it (and its counterpart of un-quantisation) evolved and was popularised throughout the past three decades. As stated earlier, un-quantisation as an aesthetic desire or preference may be seen less as Hegelian dialectical development and more a part of a wider historical and cultural epoch in which the radical traditions found in Afrodiasporic music and culture express themselves. These radical traditions stem from what Olly Wilson (1992) calls the "heterogenous sound ideal" and are materially expressed via acts of Signifyin(g) by an artist or artists.

One crucial aspect of my analysis in this chapter that may confuse the reader is the introduction of dialectical methodology for interpreting the history of quantisation and the subsequent criticism of this methodology. The reader may wonder why it would be significant to introduce and expand upon a methodology only to criticise it by the end of what is a relatively brief exploration of specific ideas and concepts. There are two purposes to this: one is to show the process for coming to the conclusion to this thesis statement: that being that the evolution of quantisation and the subsequent popularity of un-quantisation is more complex than a musical innovation created by a single individual (in this case being J Dilla). Another purpose is to show how many musicians and academics understand the role and history of quantisation, and how pertinent aspects of Western historiography determine how academics talk about musical concepts and their collective perceptions, such as how they are used or misused. Though the term dialectics tends to be used only in academic settings, the influence that it has had on Western thought cannot be understated. Even the theme of this journal, of use and misuse, revolves around the idea that musical creation can be shaped around two dialectical opposing ideas. It is easy, therefore, to fall back on this dialectical framework when discussing quantisation and un-quantisation, as its immediate impression is that of misuse.

5. Conclusions

Referring to the theme of use versus misuse, the acts of quantising and un-quantising have been thought of similarly over the course of the evolution of drum programming, with two opposing sides dialectically interacting with one another as to what will be established as correct forms of usage within hip-hop production. As this has evolved, however, what was once an impression of a misuse of technology has in some cases become an expected form of use. In my own experiences as an instrumental hip-hop producer, using various forms of un-quantisation in many circles has become expected, with implicit acts of Signifyin(g) having to do with how one presents or implements an un-quantised drum pattern within their productions. This creates a contradiction, however, in which un-quantisation becomes a trope, possibly leading to a lack of apparent Signifyin(g). In my own music, which refers to itself as lo-fi hip-hop on various streaming services, the un-quantised groove is expected by many listeners.

Within hip-hop production culture, the act of Signifyin(g) itself has close ties with the evolution of music technology and in many cases been transformed into a commodity within the hip-hop industry over the course of its historical evolution. Due to this commodification, acts of Signifyin(g) can quickly become simple tropes themselves, removed from their original musicking and simply become either a musical sign or commodity, thus turning what was originally a troping of tropes into a more straightforward trope of hip-hop production. Within a matter of just over two decades, the radical un-quantisation that was heard by the likes of Questlove and Fat Lip can now be easily accessible both in its creation by producers and its overwhelming presence in genres such as lo-fi hip-hop.

Western historiography has a major influence on how we interpret the history and culture of music technology, which, like many tenets of Western thought, may have its shortcomings. This move is a part of a wider desire to "decolonise" aspects of curricula in academia. Dialectical aspects such as use and misuse, innovation and tradition, or revolution and reaction are shaped collectively by Western cultural hegemony, which may ignore important aspects of a technology's cultural influence or roots. This is especially the case with hip-hop production and more wider music production that has been influenced by Afrodiasporic cultures among the Black Atlantic (Gilroy, 1993). Since virtually all Western popular music is in some way influenced by said cultures, methods of analysis or critique may need new ways of seeing (or in this case, hearing). In the case of quantisation, the viewpoint of this phenomena as an act of Signifyin(g) and its process into a simple trope of hip-hop production gives us a better understanding historically and culturally in its role among innovations within music production.

As established throughout this chapter, the development of both quantisation and un-quantisation within hip-hop production can be better understood as a cultural and historical process based on Afrodiasporic acts of Signifyin(g) and other practices rather than a historical framework of dialectical materialism. By viewing these acts of Signifyin(g), we can take this framework one step further by viewing how these acts are transformed through their popularisation and dissemination into commodities, resulting in their once radical act into a simple trope or characteristic specific to a genre or subgenre. Swing music, named as such because of its rhythmic characteristics contrasting to "straight" divisions of the beat found in other contemporary music genres during the height of its popularity, was once considered a radical rhythmic departure. Over the genre's evolution, however, this swing is now part of the genre itself.

As this technique of un-quantisation has further evolved into a trope, the question of misuse comes back into play, as its lack of "Signifyin(g) on the timeline", once its original purpose, becomes moot. This is seen in the criticisms of popular subgenres of instrumental hip-hop such as lo-fi hip-hop, in which the cultural roots are seen as a sort of musical gentrification, stripping away what was a basis for meaning into "muzak" for coffee shops, cafes, and student union buildings.

With the exponential growth and popularity of both the music of J Dilla and subgenres of hip-hop such as lo-fi hip-hop, the un-quantised swing that was once a radical shift has now become significantly more commonplace. One could argue that what was once a purposeful act has now become a trope in itself. Like the rhythms of jazz and their evolution in the previous century, the un-quantised grooves of this century might as well be swing.

References

Chang, J. (2005). *Can't Stop Won't Stop: A History of the Hip-Hop Generation*, New York, St. Martin's Press.

Charnas, D. (2022). *Dilla Time: The Life and Afterlife of the Hip-Hop Producer Who Reinvented Rhythm*, New York, MacMillan Publishing Company.

Charnas, D. (2023). Was Hip-Hop Really Invented 50 Years Ago?, *The Wall Street Journal* (website), available online from:www.wsj.com/articles/was-hip-hop-really-invented-50-years-ago-1dd40947 [accessed November 2023].

Danielsen, A. (2006). *Presence and Pleasure: The Funk Grooves of James Brown and Parliament*, Middletown: CT, Wesleyan University Press.

Ewell, P. (2020). Music theory and the White racial frame, *Journal of the Society for Music Theory* (website) Vol. 26, No. 2, Chicago, University of Chicago Press, available online from https://mtosmt.org/issues/mto.20.26.2/mto.20.26.2.ewell.html [accessed July 2020].

Gates H. (1988). *The Signifying Monkey: A Theory of African-American Literary Criticism*, New York, Oxford University Press.

Gilroy, P. (1993). *The Black Atlantic: Modernity and Double Consciousness*, London, Cambridge University Press.

Ferguson, J. (2014). Donuts, New York, Bloomsbury Publishing.

Fitzpatrick, R. (2011). J Dilla: the Mozart of hip-hop, The Guardian (website), available online from: www.theguardian.com/music/2011/jan/27/j-dilla- suitema-dukes. [accessed May 2023].

Floyd, S. (1997). African American modernism, signifyin(g), and Black music, in *The Power of Black Music: Interpreting its History from Africa to the United States*, New York, Oxford University Press, available online from http://dx.doi.org/10.1093/acprof:oso/9780195109757.003.0005 [accessed July 2020].

Harkins, P. (2019). Digital Sampling: The Design and Use of Music Technologies, United Kingdom, Taylor & Francis.

Harrison, N. (2004). Video explains the world's most important 6-sec drum loop (online video), available online from www.youtube.com/watch?v=5SaFTm2bcac&t=151s [accessed August 2023].

Houghton, E. (2006). The story behind some of J Dilla's greatest productions, *Fader* (website), available online from www.stonesthrow.com/news/2006/12/the-story-behind-some-of-j-dilla-s-greatest-   productions [accessed July 2019].

Hughes, T. (2003). Groove and Flow: Six Analytical Essays on the Music of Stevie Wonder, PhD dissertation, University of Washington, available online from www.steviewonder.org.uk/bio/life-stories/groove&flow/T.Hughes%20-%20Groove%20And%20Flow%20(S.Wonder).pdf. [accessed August 2021].

Reynolds, S. (2009). The cult of J Dilla, *The Guardian*, Published June 16, 2009, available online from www.theguardian.com/music/musicblog/2009/jun/16/cult-j-dilla [accessed July 5, 2018].

Robinson, C. (2020). *Black Marxism: The Making of the Black Radical Tradition*, Revised and Updated Third Edition, Chapel Hill, University of North Carolina Press.

Schloss, J. (2014). Making Beats: The Art of Sample-Based Hip-Hop, Middletown, CT, Wesleyan University Press.

Thompson, A. (2015). *Mo' Meta Blues: The World According to Questlove*, New York, Grand Central Publishing.

Wilson, O. (1992). The heterogeneous sound ideal in African-American Music, in Wright, J. (ed.) *New Perspectives in Music*, Sterling Heights: MI, Harmonie Park Press.

Discography

A Tribe Called Quest (1996) 'Get A Hold', *Beats, Rhymes, and Life*. Available at: Spotify (Accessed: 5 October 2021).

The Pharcyde (1995) 'Drop', *Labcabincalifornia*. Available at: Spotify (Accessed: 5 November 2022).

14

INNOVATION IN DANCE MUSIC RESEARCH

A Focus on Listening

Euan Pattie

1. Introduction

Throughout much writing on dance music, collective experience – be it on the club dancefloor or at outdoor raves – has been the focus of many scholars. Whilst this work highlights dance music's potential as a catalyst for building shared feelings of connectedness (Malbon, 1998), enacting political and personal freedom (O'Grady, 2015), and creating safe spaces for minority communities (Jones, 1995), by focusing almost exclusively on the collective aspect of dance music, it does not explore in-depth the significance dance music has for listeners from a more individual perspective. Indeed, listeners clearly have significant interactions with dance music outwith the dancefloor, and for some, the dancefloor is not an important site for experiencing dance music at all. The focus of this chapter, therefore, is on individual dance music listeners' perceptions of what dance music is and does for them; how listeners interact with and use dance music throughout their lives, without necessarily focusing on the dancefloor. Through analysing findings from in-depth semi-structured interviews with dance music listeners, a number of conclusions are highlighted, based on participants' reported perceptions, feelings, and experiences. Firstly, it is highlighted that dance music is a complex object of study and questions are posed as to the extent engagement with dance music is influenced by the music and its general connotations, or the tendencies of individual listeners. This questioning is continued in outlining that listeners of dance music have complex and often contradictory attitudes towards drug use and collective dance music experiences overall. Finally, attention returns to more individual experiences of dance music, specifically on how listeners engage with various analogue and digital technologies, where various positive and negative aspects of streaming services are highlighted.

Here, 'dance music' is taken as an umbrella term encompassing several categories of music including house, techno, garage, and disco (Fikentscher, 2000), excluding other types of 'dance music' such as ballroom styles, tango, or Charleston. Whilst debates over what may be considered 'authentic' dance music – such as the hardcore continuum debate (Reynolds, 2010) – involve complex historical and cultural factors, in the spirit of keeping this chapter's focus closely on the perception of individual participants themselves, the definition of dance music has been left largely open to interpretation by the participants.

DOI: 10.4324/9781003396550-14

2. Dance Music's Impact on Listeners

In exploring dance music's impact on listeners, DeNora's (2000) conception of music as functional – that music is a potential resource people may draw from in various ways in different scenarios – may be employed. By focusing on music's functionality, we may begin to understand more music's effects on individuals in social life, rather than viewing music strictly as an object from which meaning can be directly 'decoded'. Schäfer et al. (2013) usefully summarise potential functions of music under three categories: to regulate arousal and mood, to achieve self-awareness, and as an expression of social relatedness. These categories are explored further below, helping to form the basis for this study's interview questions.

Firstly, music's capacity to arouse feelings is frequently cited (Juslin and Sloboda, 2010). DeNora (2000), for instance, suggests people often use music to attain and/or maintain states of feeling in different scenarios, often linked to achieving certain goals – e.g., to help one concentrate, relax, or feel energised. On music's specific emotional impact, scholars from a range of disciplines contest numerous issues, making dealings with music and emotion difficult to discuss with much accuracy or consensus. For example, within music psychology, scholars contest whether music should be considered able to truly induce, as opposed to merely express, emotions (Juslin and Laukka, 2004), whereas some music philosophers doubt whether music should be considered expressive of emotions at all (Davies, 2010). Considering that most academic theories relating to emotion are likely unclear to most listeners, the focus of the current study is more on how listeners themselves construct and articulate their feelings about music, to which academic concepts relating to emotion can be applied to responses after. Secondly, functions pertaining to self-awareness involve using music to help "people think about who they are, who they would like to be, and how to cut their own path" (Schäfer et al., 2013; p. 6). Here, DeNora (2000) proposes that music may be used to build a sense of the 'self for the self', through helping listeners remember memories and 'turn over' past experiences; in some ways, to help listeners remember who they are. Music's capacity to help build and recall memories is significant and widely cited, with many scholars noting how specific pieces of music become strongly associated with particular times in individuals' lives (Juslin and Laukka, 2004). Finally, music's uses in social relatedness may be the most widely discussed by scholars of Schäfer et al.'s (2013) three categories, with frequent reference made towards music's role in formulating conceptions of self-identity. Goodman (2011) notes the importance of music in 'self-making', arguing against the romantic attitude that the appreciation of music is a solely inward, introspective, and emotional venture, whilst Frith (1996) and DeNora (2000) suggest people 'try on' new identities using different music in a similar manner to clothes – that rather than being fixed, self-identity is a product of social 'work' put in by the individual.

In exploring music's impact on listeners, it is furthermore important to consider how individual listeners specifically engage with music and the impact of different listening contexts/ecologies. Various types – or modes – of listening have been conceptualised, including Adorno's (1938) regressive listening, Schaeffer's (1966) reduced listening, Demers' (2010) aesthetic listening, and Oliveros' (2015) deep listening, each relating to the level of attentiveness given towards the music. For Adorno, distracted/inattentive listening is seen as regressive, reflective of the impoverished and superficial nature of popular music. Demers, however, views distracted listening more positively, emphasising that individuals have much to gain from such 'listening with feeling' alongside other tasks. Listener attentiveness may also be impacted by the listening setting and choice of playback technology, in terms of playback systems (headphones, sound systems, phone/laptop speakers), musical format (vinyl, tape cassettes, CDs, digital MP3/WAV, bought/streamed), and how the music is organised (whether by albums, singles, artists, or playlists).

With these points on general music listening in mind, Frith's (1996) argument that genre conventions affect how listeners interpret, appreciate, and interact with music emphasises a need to examine what is specific about dance music that it may be listened to in any typical way. Although dance music may be considered a metagenre comprising of as many as 300 subgenres (McLeod, 2001), much of this music largely shares certain aesthetic and cultural connotations – for example, its longstanding associations with club/sound system culture (Reynolds, 2013; Henriques, 2011), dancing itself (Fikentscher, 2000), drug use (Collin, 2010), and crate digging for 12-inch vinyl singles (Hesmondhalgh, 1998). Therefore, it is significant to explore whether and how these associations impact how listeners interact with dance music throughout their everyday lives. For example, how might listeners engage their body when listening to dance music in more everyday settings and how do they feel this impacts what the music does for them? How may listeners perceive the effects of drugs or alcohol on experiences of listening to dance music?

3. Getting at Listeners' Perceptions, Feelings, and Experiences

As the discussion of literature above demonstrates a focus on music listening more personal to individuals (emotion, self-identity, experiences, attitudes), rather than specific to any collective movement or culture, in-depth semi-structured interviews with individual listeners of dance music was chosen as the primary research method over any distinctively ethnographic approach. A total of 18 interviews were conducted (lasting between 40 and 95 minutes), using questions deriving from the exploration of literature above. Participants were recruited using the broad inclusion criteria of any self-reported regular listener of dance music over the age of 18, leading to the recruitment of participants of a variety of ages (18–60+), genders (3 female and 15 male), and musical experiences/ preferences. With a relatively small participant group size of 18, there was no attempt made to 'represent' dance music listeners demographically in terms of ethnicity, gender, social background, sexuality, etc. To make such an attempt – especially when these groupings do not directly relate to the research questions – may be viewed as tokenistic and might run the risk that unfair extrapolations pertaining to these demographic groupings be made from limited data. While being cautious in assuming everything said by interviewees is entirely representative of their experience, the analysis of data is grounded in Hesmondhalgh's (2007) advice of taking what interviewees say as having at least some degree of meaning. Moreover, it is widely recognised in the methods literature that despite their limitations, well-conducted in-depth individual interviews have great potential to provide fruitful data on individuals' perceptions, feelings, and experiences (Schutt, 2001).

Reflexive thematic analysis was chosen as the approach to data analysis, since its highly fluid and exploratory nature allows for a broad number of topics to be explored and for flexibility in the pieces of data focused on as the analysis develops (Braun and Clarke, 2022), qualities suiting this study's broad, exploratory research areas. Whilst the findings presented here are only some taken from a wider study (in which coding and theme development was still ongoing at the time of this chapter's production), what is analysed in the following discussion are participant responses relating to three initial patterns of meaning in the dataset: what dance music is and what it does, attitudes towards drug use, and playback technologies. Whilst it is not the intention that these findings be representative of all dance music listeners, there is value here in exploring the reasoning behind relevant participant responses, the nuances between similar and differing responses, and how these responses relate to existing literature and normative thinking on the topics highlighted.

4. What Dance Music is and What It Does

Firstly, it is clear from the initial reading of the data that participants demonstrate differing presumptions about the essence of dance music as a whole. For some, dance music is distinctly uplifting and joyful; for others, it is distinctly dark and moody:

> I think mostly dance music is feel-good music. There's not a lot of dance music that's sombre or makes you feel shitty or is about heartbreak necessarily. It's all catered towards getting people moving.
>
> *Participant 3*

> It's never angry . . . I don't think I've ever heard a dance song in my life that I've been like, 'That's actually, like, actively aggressive'. The whole point, obviously, is to leave all that at the door.
>
> *Participant 8*

> If it's dance music, you don't turn the lights on, you turn the lights down, you know? It's always music you'd put on at the end of the gaff. Or you grow towards the dance music, usually, 'cos it's darker, it's more rowdy.
>
> *Participant 6*

That 'dance music' be interpreted in a variety of ways appears at least partially due to the fact that what is being discussed here is a metagenre encompassing a wide range of music of different sounds, themes, and moods. Whereas Participant 3 argues dance music is rarely sombre, other listeners may argue the likes of Burial's dystopian garage is very sombre dance music, whilst Participant 8's argument that dance music is never aggressive could be countered with reference to music of the likes of The Prodigy. Although 'dance music' generally denotes certain cultural and aesthetic connotations (e.g., dancing, collectivity), these appear to only apply to all the music that may be considered dance music to varying degrees (e.g., the controversially named I.D.M.), and so to pinpoint a universal essence of dance music is a complex – if not impossible – task. It appears, then, that individual listeners form personal conceptualisations of what dance music as a whole is, likely influenced in part by general connotations and by individual experience and taste – by the specific 'dance music' they have listened to and how they perceive that music. That participants outline differing conceptualisations of dance music highlights the need to take what listeners say about dance music in the context of individual perceptions, and that research focusing on dance music as a whole may produce complex results.

Keeping with the notion that perceptions of dance music are personal to individuals, participants describe why they like the dance music they listen to using a variety of different types of responses, including those focusing on mood regulation, physiological affect, cognitive analysis, and self-identity. Participant 7, for instance, reflects DeNora's (2000) notion that music is a potential resource for aligning one's mood with the demands of ongoing goals, noting she uses dance music to relax:

> First and foremost, I use music to decompress. And what I mean by that is, work gets busy and stressful, and it just takes up so much of my time and energy and thought process. It is my go-to decompression activity.
>
> *Participant 7*

Others reference more physiological effects of dance music on the body:

> So, physical response, rather than emotional. But of course, you get an emotional response to the experience but in terms of the actual music itself, for me, not so much, you know? More of a bootie response rather than a cerebral response.
>
> *Participant 13*

Some participants engage with dance music in a more analytical manner, referencing musical elements such as rhythm, melody, harmony, and production:

> I like tracks that use time signatures other than 4/4... There's a few tracks where you've got the drums that are in 4/4 and then the synths are in 5/4, or something. And then that... intersection between the two time-signatures is quite interesting because they overlap and sync up at places ... It can be quite hypnotic.
>
> *Participant 10*

Others note how dance music can reflect personal political ideologies, influencing self-identity:

> So, for me, that goes absolutely to the core of my ethos on life. Love your fellow humans, respect everybody's right to individuality, individual choice, freedom, freedom of expression.
>
> *Participant 13*

These responses highlight the multifaceted ways in which dance music as a whole is significant to listeners. Whilst some responses describe engaging with the music in a distinctly somatic sense (dancing), others describe far more cerebral engagement (using dance music to relax, analysing the music). Again, this demonstrates that listeners have much to gain from engaging with dance music in ways that do not align with the metagenre's general connotations of dancing and collectivity. Furthermore, individual participants do not exclusively refer to one type of engagement – Participant 13 notes that whilst he enjoys the somatic experience, it is also important that his ideological beliefs are reflected by the music. It is clear, then, that there are many – often contrasting – ways in which listeners relate to dance music and individual listeners may garner significant meaning from dance music in multiple ways simultaneously. This range of responses may again be partially explained by the fact the focus here is on dance music as a whole, not a more specific type of music, and so further research could focus on listeners' relationships with more specific types of dance music or whether there is any correlation between engagement and type of dance music. Whilst out of the scope of this chapter, these points highlight questions on determinism, useful for exploring listeners' perceptions and engagement with music. For example, to what extent do listeners engage with the different types of dance music in ways implied by the music itself and its connotations, as opposed to other factors such as setting and their own agency to engage with – and repurpose – music in any way they desire?

Whilst participants clearly articulate uses of dance music in the ways noted above, notably more complex to discuss is the allusive nature of emotional affect. When questioned on dance music's emotional impact, some participants report not consciously thinking in terms of emotion when listening, instead refocusing their responses on other ways dance music is significant to them:

> I think it is a hard one because ... I don't really, when I'm listening to music, really think too much about it. It's just something that goes on in the background ... I kinda know what it's

gonna make me feel but I'm not really paying attention to that. But I think definitely it has a soothing effect on me.

Participant 17

In understanding emotional affect further, music psychology scholars Juslin and Sloboda (2010) provide a useful distinction between mood and emotion – whilst mood may be thought of as a long-lasting, yet low-in-intensity affective state, emotion is far more intense, yet shorter lasting. Schäfer et al. (2013) also suggest that listening to music for emotional uses requires conscious effort and deep attention, whereas listening to attain a pleasant mood does not require deep involvement with the music. Here, Participant 17 argues that although he does not focus directly on any emotional impact when listening, he concedes he does feel a 'soothing effect' from the music. It appears, therefore, that what he is referring to is using dance music more for mood regulation – 'in the background' – than to feel any intense emotional affect. The effect of this background listening on his mood also appears to be perceived more as a positive after effect, than a conscious goal from the onset. These points call to attention questions relating to listener self-awareness and the extent to which listeners themselves feel they use dance music consciously in the ways outlined by the likes of DeNora (2000). For example, whilst the two may not be considered exclusively separate, to what extent do listeners listen to music conscious of its functionality – with the expectation that 'X music has X effect on me' – as opposed to listening to music simply because they appreciate its aesthetics? Although there is limited space here to address these issues fully, they are significant in exploring how listeners themselves perceive the role music has in their lives.

5. Attitudes Towards Drug Use

Continuing to address how common connotations of dance music influence listener perceptions, participant data highlights strong opinions on the role of drugs in listening to dance music. Firstly, there is evidence across the dataset of participants making a distinction between listeners either using drugs to 'enhance the music' or to 'enhance themselves as individuals':

I always got that level of, it enhanced the music, rather than enhancing you individually.

Participant 2

I think maybe having a wee bit of something makes the music seem a bit smoother.

Participant 11

Drugs help you get into, like, a body state. It's almost like the drugs help you get your body into the same state as the music.

Participant 8

Participant 2's response demonstrates a clear distinction between listeners using drugs to alter their perception of the music and listeners using drugs to alter themselves as individuals, whilst Participant 11 exclusively describes using drugs to enhance perceptions of the music. Participant 8's response describes a mixture of the two – that drugs may be used to enhance the individual, as a means to enhance the perception of the music.

That participants suggest some people use drugs primarily to 'enhance themselves as individuals' implies that these people use drugs more to feel – and enjoy – the physical and mental effects of the drugs, than as a means of specifically improving the listening experience, through reaching an 'understanding' of the music in a sense not possible sober. That participants distinguish these motivations for using drugs alludes to an underlying bias perpetuated by many – that some listeners take drugs to enjoy the music more, which is considered commendable, whilst others use the association dance music has with drugs as an excuse simply to take drugs to enjoy their physical and mental effects, which is considered selfish. Indeed, many echo that:

A lot of people do go to raves just to take drugs because that's where it's socially acceptable.

Participant 1

Phrases such as 'just to take drugs' allude to an implied hierarchy of superiority of listener – that those who take drugs specifically to enjoy the music more are seen as truer to the music than those who take drugs simply to enjoy their effects. It is significant that these participants are not entirely against taking drugs, yet they do stigmatise others who take drugs for the supposed 'wrong reasons'. This attitude is not limited to illegal drugs, some showing similar dismay towards those who focus more on drinking alcohol for its inebriating effects on a night out, than on enjoying the music:

Some of my friends are more about the drinking and getting absolutely messed up rather than going to actually dance about in Bongo's, d'you know what I mean? So sometimes they just want to sit in Opium and get pissed and it's like, 'This is so bad!'

Participant 12

Other participants appear to be against drug taking for any reason, proudly stating they do not 'need' drugs to enjoy the music:

I think it depends on the person and your mindset and your personality type. Some people, you know, find it difficult to enjoy it and engage with it without that sort of enhancement . . . Personally, I don't really see the need for it.

Participant 7

That participants argue they do not need to resort to taking drugs to enjoy dance music implies the attitude that they have a closeness and trueness to the music that others would never achieve sober. However, it seems these participants place far more value on appreciating the music itself – as if it exists in a vacuum – than on the overall experience of listening to dance music and the impact of setting. Others do appear to value the overall experience more, noting that club scenarios offer a way to appreciate dance music in a different light:

There's definitely some music . . . techno, for example, that I would only listen to in a club setting, usually under the influence of drugs of some kind. I wouldn't listen to techno in the house, and I'd probably actually find it quite unbearable to listen to in that setting . . . But in that club setting, under the influence, with friends, it's great.

Participant 18

That some participants have such clear biases against drug taking highlights how dance music's association with drugs – perhaps even the values of hedonism and freedom – are only relevant to listeners of dance music to varying degrees. An important factor here is clearly space, in that clubs are viewed by many participants as natural – perhaps even acceptable – places to take drugs. Indeed, Participant 18 outlines how the combination of the club setting and drug use allows for a different engagement and appreciation of music she would otherwise find 'unbearable' at home. Whilst setting falls under the broad topic of wider listening ecologies too vast to be fully addressed in this chapter, the next section turns attention towards how listeners engage and interact with dance music away from the club, specifically on the impact of various analogue and digital playback technologies.

6. Playback Technologies

Whilst many participants describe how digital technologies such as streaming services have become embedded in their listening practices, they also note that the emergence of these technologies has not led to a complete abandonment of analogue playback such as vinyl and CDs, calling to attention how listeners may use older and newer technologies simultaneously. Outlining her views on how vinyl and Spotify listening experiences differ, one participant notes:

> There's something special about vinyl . . . just sort of takes that bit more time and effort to play those songs. And usually it involves listening through an album if it's an album. Or if it's a single, even just having to pick it up and switch it round and really be present with the piece of music and making an effort to source it and . . . you're careful with it and you're making sure there's no scratches and there's something really special about that and it being a bit more of a slower process . . . I think Spotify is a good way to explore and discover music. A lot of the times you'll play a song and then you'll get your recommended songs and it'll sort of play in order and you can fast forward and skip and delve into a particular thing a bit more. But I think there's something also a bit less genuine about it. Because a lot of the time, I'm listening to music and . . . I don't take note of who it is or what song it is and I'll maybe never be able to find that song again.
>
> *Participant 18*

Many participants echo these sentiments, noting whilst streaming services offer a way to discover and listen to new music in seamless automatic sequences, the convenience of this affects the sincerity of the listening experience. Many argue that listeners often give the music less attention when played through streaming services, skipping through songs, or not taking note of the artist playing, contrasted with the care and attention involved in the 'slower process' of vinyl listening. However, it is important to note that the dataset does not reflect a complete dichotomy between streaming services resulting in inattentive listening and analogue devices resulting in fully attentive listening. Many participants do report listening attentively via streaming services and claim that the value they give towards the music is unaffected by playback method. This again leads to questions on determinism, specifically the extent to which playback methods imply certain modes of engagement with music that are generally followed by listeners, as opposed to other factors such as the tendencies/agency of individual listeners. Moreover, that these responses place clear emphasis on attentiveness alludes to an Adorno-like superiority over distracted listening, also perpetuated elsewhere in the dataset:

> (I) genuinely feel that the way people listen to music now is not very good and the lack of concentration span when they're doing it, people mostly . . . listen to music, you know, commuting, out walking, maybe have it on in the background but they're not really giving it the attention that it really deserves.
>
> *Participant 2*

That some participants assume dance music should always be listened to with full attention appears strange considering the clear connotations dance music has with dancing and collectiveness, in addition to suggestions from other participants that dance music should always be danced to. This again demonstrates that attitudes towards how dance music 'ought' to be listened to differ vastly between listeners and do not always correspond with the metagenre's general connotations.

Furthermore, it is significant that listeners do not purchase the music they listen to via streaming services, instead paying a subscription to access all songs available on the specific site. Overall, participants demonstrate differing attitudes towards how this impacts the value they give towards the music, some suggesting that not purchasing music has resulted in them valuing the music less:

I: Do you think that (not buying the music) impacts on how you value the music?
P: 100%. Yeah. Massively. I think it kind of cheapens everything . . . just because it's so easily
      attainable . . . Something about everything being available takes away a little bit of the magic
      of finding things. Finding tracks. Especially if you're DJing, I think a lot of the joy of it and
      the art of it comes from digging and knowing where to look for good records . . . Whereas
      I feel like now that doesn't really exist anymore. Or if you do it . . . you kind of make it hard
      on yourself.
                                                                                Participant 9

Here, Participant 9 again describes how the convenience of being able to listen to vast amounts of music for a relatively small subscription fee 'cheapens' the overall experience, through taking away the 'magic' of finding music in places one needs to have specialist knowledge of. He argues that whilst listeners may still find music in this way, it is now not the normative practice, with streaming services offering far more convenience. This evidences a clear departure from the crate digging of 12-inch vinyl singles prominent in the 1980s and 1990s, highlighting a need to further explore how the culture of 'digging' has changed with the development of digital technology. Other participants, however, argue that not buying music directly does not impact on how they value the music itself:

    If I like it, it's valuable to me. And money-wise . . . there's no monetary value in it at all, is
    there? . . . Spotify and YouTube and all that have kind of spoiled that for the artist, which is a
    shame . . . No additional value if I paid for it or if I didn't.
                                                                                Participant 11

Whilst Participant 11 argues streaming services are likely damaging to the careers of artists, he appears to solely blame the streaming services themselves for this, rather than accept any blame for his part as a user. Although other participants note they attempt to support the artists they listen to through purchasing their music, it is significant here that 'free' music is not a phenomenon that has emerged solely with the advent of streaming services:

    In the 1900s, we used to just put everything on cassettes, so . . . I had tonnes of albums that I
    never bought . . . I don't think that there was ever a time in my life where I really thought, 'Oh,
    I have to support this artist by buying their music' . . . So, I think it would be a bit unfair of me
    to sit and say, 'Oh yes, terrible, these kids getting everything for free'.
                                                                                Participant 13

These points highlight that although listening has changed dramatically in the past three decades, aspects of pre-digital listening remain, and many listeners today use a mixture of analogue and digital playback technologies. Whilst the dataset does not demonstrate any correlation between reported value given towards the music and playback technology/whether music is bought or not, this again emphasises the tendencies of individual listeners in perceptions of music. Overall, however, participants argue that despite the negative aspects of streaming services, their convenience and affordability is enticing enough to make them the preferred format for many listeners.

7. Conclusions

This chapter features analysis of findings from in-depth semi-structured interviews of dance music listeners, exploring how listeners interpret their experiences with dance music. Although the ideas explored here are part of a wider study still in its developing stages, a number of conclusions are highlighted from this initial analysis of findings. Firstly, that 'dance music' comprises a wide range of music of different sounds, themes, and moods makes for a complex object of study, where participants demonstrate differing presumptions and uses for dance music itself. This leads to questioning the extent to which engagement with dance music is influenced by the music and its general connotations, or other factors such as setting, or the tendencies/agency of individual listeners. That some participants do not consciously think in terms of emotion when listening to dance music also highlights interest for further study in exploring the extent to which listeners feel they listen to music for more functional, as opposed to simply aesthetic, reasons. Further addressing how common connotations of dance music influence listener perceptions, participants imply a hierarchy of superiority of listener in terms of drug use, where those who take drugs to enjoy the music more are often seen as truer to the music than those who take drugs supposedly to simply enjoy their effects. Others proudly state they do not need drugs to enjoy the music at all, appearing to place far more value on appreciating the music itself than on overall experiences of dance music. From this it is clear that dance music's association with drugs – and perhaps even the values of hedonism and freedom – are only relevant to dance music listeners to varying degrees. Finally, participants describe how various analogue and digital playback technologies involved in listening impact on their interaction and appreciation of dance music. Whilst the dataset does not demonstrate any correlation between reported value given towards the music and playback technology/whether music is bought or not, participants overall suggest it is the convenience of streaming services that make them the preferred format for many. Whilst conclusions from this study are currently still developing, this chapter highlights an interest in less prominently researched facets of dance music listening and raises questions relevant to further studies on listening of any genre.

References

Adorno, T. W. (1938) 'On the fetish-character in music and the regression of listening', *Popular Music: Critical Concepts in Media and Cultural Studies*, 3, pp. 325–49.

Braun, V. and Clarke, V. (2022) *Thematic analysis: a practical guide*. Los Angeles: SAGE.

Collin, M. (2010) *Altered state: the story of ecstasy culture and acid house*. London: Profile Books.

Davies, S. (2010) 'Emotions expressed and aroused by music; philosophical perspectives', in Juslin, P. N. and Sloboda, J. A. (eds.) *Handbook of music and emotion: theory, research, applications*. Oxford: Oxford University Press.

Demers, J. T. (2010) *Listening through the noise: the aesthetics of experimental electronic music*. Oxford: Oxford University Press.

DeNora, T. (2000) *Music in everyday life*. Cambridge: Cambridge University Press.

Fikentscher, K. (2000) *"You better work!": underground dance music in New York*. Middletown, Conn.: Wesleyan University Press.

Frith, S. (1996) *Performing rites: evaluating popular music*. Oxford: Oxford University Press.

Goodman, D. (2011) 'Distracted listening: on not making sound choices in the 1930s', in *Sound in the age of mechanical reproduction*. Philadelphia: University of Pennsylvania Press. pp. 15–46.

Henriques, J. (2011) *Sonic bodies: reggae sound systems, performance techniques, and ways of knowing*. New York; London: Continuum.

Hesmondhalgh, D. (1998) 'The British dance music industry: a case study of independent cultural production', *British Journal of Sociology*, 49(2), pp. 234–251.

Hesmondhalgh, D. (2007) 'Audiences and everyday aesthetics: talking about good and bad music', *European Journal of Cultural Studies*, 10(4), pp. 507–527.

Jones, S. (1995) 'Rocking the house: sound system cultures and the politics of space', *Journal of Popular Music Studies*, 7, pp. 1–24.

Juslin, P. N. and Laukka, P. (2004) 'Expression, perception, and induction of musical emotions: a review and a questionnaire study of everyday listening', *Journal of New Music Research*, 33(3), pp. 217–238.

Juslin, P. N. and Sloboda, J. A. (2010) *Handbook of Music and Emotion: Theory, Research, Applications*. Oxford: Oxford University Press.

Malbon, B. (1998) 'The club: clubbing: consumption, identity and the spatial practices of every-night life', in *Cool Places: Geographies of Youth Cultures*. London: Routledge. pp. 266–86.

McLeod, K. (2001) 'Genres, subgenres, sub-subgenres and more: musical and social differentiation within electronic/dance music communities', *Journal of Popular Music Studies*, 13(1), pp. 59–75.

O'Grady, A. (2015) 'Dancing outdoors: DiY ethics and democratized practices of well-being on the UK alternative festival circuit', *Dancecult*, 7(1), pp.76–96.

Oliveros, P. (2015) *The Difference between Hearing and Listening | Pauline Oliveros | TEDxIndianapolis*. Available at: https://youtube.com/watch?v=_QHfOuRrJB8&feature=share (Accessed: 10.01.22).

Reynolds, S. (2010) 'The history of our world: the hardcore continuum debate', *Dancecult*, 1(2), pp. 69–76.

Reynolds, S. (2013) *Energy flash: a journey through rave music and dance culture*. Rev. ed. London: Faber and Faber.

Schaeffer, P. (1966) *Traité des objets musicaux*. Paris: Le Seuil.

Schäfer, T., Sedlmeier, P., Städtler, C. and Huron, D. (2013) 'The psychological functions of music listening', *Frontiers in Psychology*, 4, pp. 511.

Schutt, R. K. (2001) *Investigating the social world*. 3rd ed. Thousand Oaks, CA: Pine Forge Press.

15

CREATIVE CYBORGS

Virtual 3D Characters as Artist Identities for Musicians

Kirsten Hermes

1. Introduction

This paper explores how motion-controlled virtual avatars can be used as artistic identities for musicians. The democratization of audio technology and internet file sharing have turned music production from a craft for the few into a hobby for many. Affordable DAWs allow anyone to record, arrange, edit, mix, and master music at home. Similarly, free graphics software tools make the creation of 3D graphics possible for the masses. Still, the creation and animation of unique virtual characters requires skills that lie outside of the typical area of expertise of DIY artist-producers.

The paper assesses the creative challenges and affordances of creating motion-controlled characters through a DIY approach. Existing case studies and popular software tools are investigated, and an experiment is carried out, whereby the author creates a virtual doll in the 3D graphics software Blender (2023) and uses Mixamo (2023) for animation. In addition, movement from a real-live dance performance is applied through Plask, an AI-based motion capture tool. Initially centred around a music video, this exploration also considers applications in live performances. Section 2 reviews the work of other musicians. Section 3 introduces the case study, presenting a technical breakdown of the methods employed in this study, followed by insights into the creative challenges encountered throughout the process. Csikszentmihalyi's (2013) flow theory is used as a framework for this investigation. Section 4 discusses the findings and Section 5 concludes.

2. Virtual Characters and Purposeful Artifice

Why do musicians work with avatars? According to Auslander (2009, pp. 303–316), Musicians do not only create music but enact roles. In a competitive market, standout personas attract followings (Gladwell, 2002; Godin, 2005). Clear and consistent signs, semiotics, and core values forge strong artist-fan connections (Morris, 1971). Since visual technologies became more commercialized in the latter half of the 20th century, artists and their labels began to purposefully fashion marketable characters (Diederichsen, 2015). For example, the Beach Boys constructed their personal narratives within music videos through a tapestry of sun-kissed adventures and harmonious bonds that became synonymous with their iconic sound (Diederichsen, 2015). Some modern artists, such as Marshmello, Deadmau5, and Sia even adopt props to conceal and alter their identities, adding

DOI: 10.4324/9781003396550-15

an air of intrigue. Marshmello and Deadmau5 are both known for their characteristic helmets, and Sia uses oversized wigs, to shift the focus from real-life personas to the music and craft itself. Edlom (2020) asserts that authenticity in pop music is deliberately crafted to foster genuine audience bonds. Peterson (1997) notes that this perceived authenticity is fabricated and subjective, and Middleton (1990) even writes of a "debris of authenticity". Born and Hesmondhalgh (2000) declare the subject of authenticity to the "intellectual dust heap". Frith (2001) asserts that as long as the "magic is real", it does not matter whether it is constructed or not. Virtual identities appear to be a natural progression from the above examples, using modern software to construct artist identities. Section 2.1 introduces artist personas that are constructed through digital technologies, and section 3.1 focusses AI specifically. Section 3.1 reviews the options for DIY musicians to create their own custom avatars.

2.1. Digital Artist Personas

The concept of constructed authenticity is taken to the extreme as artists coalesce with, or are entirely replaced by virtual characters. Here, technology is elevated from a silent contributor to an active collaborator. ABBA's London show "ABBA Voyage" sees the four musicians relive their '70s heyday as de-aged digital avatars in a highly realistic CGI show (ABBA Voyage, 2022; Plaete et al., 2022). Through an immersive reality concert produced by Sony Immersive Music Studios, Magnopus, Gauge Theory Creative, and Hyperreal (2021), Madison Beer was able to perform to fans in VR during the pandemic, as a 3D avatar resembling her in real life. Transgender artist Sophie was renowned for using plastic surgery and digital technology to shape her distinct and personalized identity, offering new forms of expression to her transgender fans (Rovinelli, 2021). In doing so, she elevated the process of identity crafting to become a form of identity in itself: the Acid Horizon podcast (2021) presents this process as a "remanufacturing of the self": "there is no authentic self to be presented, so there is nothing but the presentation". The point is to construct a complete artifice, to bring to light the artificiality of identity.

Not all virtual avatars are representative of real human creators. An illustrative example is Hatsune Miku, a virtual pop sensation originating from Crypton Future Media's Vocaloid software (2023, a music technology company based in Japan), introduced in 2007. Vocaloid is a type of voice synthesis allowing users to craft and manipulate singing voices, and eliminating the necessity for human vocalists. Within the framework of the software's branding, Crypton Future Media introduced a series of anime-style characters, with Hatsune Miku emerging as a standout figure. Recognized for her distinctive turquoise hair and computer-generated voice, she has found her way into music, concerts, rhythm games, and more. Her deadpan facial expressions and deliberately unrealistic features have endeared her to a passionate fan base. Through holographic performances alongside live musicians, Hatsune Miku also "performs" on stages worldwide (Hermes, 2022). The multiplayer online video game League of Legends (LoL) used imaginary girl group K/DA as part of its promotional strategy, consisting of four female stars that are also avatars in the game. K/DA's live shows bring together real and fictional musicians (Huttu-Hiltunen, 2020). The Gorillaz are an English virtual band consisting of fictional cartoon characters.

2.2. Creative Cyborgs: AI Co-creativity

Many artists collaborate with AI (Artificial Intelligence), to create virtual characters, or use the appeal of AI as part of the brand. AI refers to machines learning from experience and adjusting to new inputs, instead of being explicitly coded. Within AI, the field of machine learning enables computers to carry out tasks which are non-routine and require creativity (Susskind, 2020). AI can be seen as a creative actor through its intriguing capacity to emulate human skills and simulation of divergent thinking (de Mántaras, 2020; Dautenhahn, 2007). In creating her digital twin Holly+ (2023), Holly Herndon has not only crafted an AI-powered vocal deepfake of her voice, but an expanded artistic identity where Holly+ is visually represented as a digitalized version of the artist herself.

The fascination with AI can also be a powerful marketing tool. Lil Miquela (Lord, 2018) is a fictional digital character and virtual influencer created by the company Brud, designed to appear as an "AI robot" resembling a 19-year-old Brazilian-American model and musician. She showcases a mix of fashion, lifestyle content, and even music releases, and interacts with her followers on social media platforms, as if she were a real person. Her creators have crafted a detailed backstory for her, and she often engages in collaborations with real-world brands. Lord (2018) posits that we are teaching Lil Miquela the contrived affectation of social media: "Ironically the faker Lil Miquela acts – pouting into glamorous bathroom mirrors, posing in front of impossibly large brunch spreads – the more she becomes like us."

Most of the above virtual characters appear purposefully artificial, and their artificiality is a core aspect of the branding strategy. The method of their creation, using cutting-edge technology, is an overt, and very visible part of the creative concept. For example, Lil Miquela refers to herself as an "AI", or a "robot", and Hatsune Miku is very clearly a game character with unrealistic features. Part of the appeal of "ABBA Voyage" is the excitement around experiencing impressive new technologies live. Some artists, such as Holly Herndon, along with their virtual representations, become creative cyborgs, where the collaboration with a technological "other" (usually AI) is part of both the creative concept and the method of its creation. As in the case of Madison Beer, new technologies allow fans to be closer to their favourite artists, and to experience their performances in new ways. The creative collaboration between human and machine is reminiscent of Deleuze and Guattari's assemblage theory (1988), as well as Donna Haraway's feminist posthumanist "Cyborg Manifesto" (1991), which rejected the rigid boundaries between human and animal, and human and machine. Here, the concept of a cyborg represents the plasticity of identity and highlights the limitations of socially imposed identities.

Our fascination with sentient machines has a long history, dating back to ancient myths and folklore and finds itself as a repeated theme in Sci Fi literature (Ajeesh and Rukmini, 2023). In reality, while AI tools may produce unexpected and novel outputs, they cannot move beyond their own intended purpose, simulate human thinking and reasoning, or act independently. AI also cannot combine creative skills like humans can (Hertzmann, 2018, p. 19), or conceptualize goals or inputs meaningfully (Gioti, 2020, p. 30). Instead, AI can be a complementing force for human work (Susskind, 2020). To describe this blended creativity, the concept of "co-creativity" has emerged (Wingström et al., 2022). "Augmented intelligence" (Carroll, 2021) is an alternative conceptualization of artificial intelligence (AI) that focusses on AI's assistive role in advancing human capabilities. The complex interaction between AI and human creatives is in line with the notion of the "creative cyborg" discussed here, and reflects how artists also visually "collaborate" with their virtual avatars, creating an air of mystery and technofetishism.

2.3. Options for DIY Artists

Role playing appears to be a natural human inclination, exemplified through activities like cosplay, role-playing games (RPGs), theatrical performances, or dressing up. But how can DIY musicians become creative cyborgs? Many social media platforms, such as TikTok (2023), offer free editing capabilities, while AI-driven phone applications like Facetune (2023) facilitate virtual makeup application and facial alterations, even for individuals with minimal digital editing proficiency. Within the metaverse and virtual reality (VR) domains, such as VRChat (2023) or the Meta-affiliated Horizons (2023), players can socialize as their virtually constructed counterparts. VTubers harness anime-inspired characters for talking-head videos. Often based on anime aesthetics, these characters can be created in platforms such as VRoid Studio (2023) or VTuber Maker (2023) and can be animated using motion capture facilities built into readily available consumer hardware (facial motion capture is possible on iPad, for example). More complex movement sequences, such as dance, are more difficult to implement, however. On a simpler level, in video games, or desktop chat platforms like IMVU (2023), players can design and customize characters and buy new clothes for them using in-game currencies, and many of these platforms allow players to export their characters for use in other contexts.

While most consumer-facing platforms are intuitive and easy to use, creators with distinct and specialized visions may encounter creative bottlenecks when venturing beyond the confines of these platforms' established aesthetics (and restricted mocap functionalities). Characters designed in video games are perceived as playful interactions with these games, and are less respected as standalone artistic outputs. For a uniquely defined custom character, delving deeper and creating assets from scratch becomes imperative, albeit notably more challenging. To animate characters for music videos or on stage can also be difficult, requiring rigging, as further explained in section 3. Artists can either program movement sequences manually in 3D software tools or games engines (which is a slow and complicated process), or use motion capture, whereby a real human's movements are captured and transferred to the character. Motion capture traditionally requires specialist teams and suits or trackers, especially for complex movement sequences, such as dancing, though cheaper, consumer-facing options are emerging (such as Sony's Mocopi). A few AI tools have also become available (such as Plask, as introduced in section 3). The aim, therefore, is to answer the research question: how accessible is the process of creating custom virtual characters to DIY musicians, through a DIY (do-it-yourself) approach?

The options include free game engines like Unity (2023) and Unreal Engine (2023), where characters can be created and animated in many ways. However, this requires intricate knowledge and a steep learning curve. Similarly, 3D graphics programs like Blender (2023) require advanced knowledge in modelling, rigging and animation, as explored further in the next section. There are simplified character creation tools, such as Ready Player Me (2023), a cross-game avatar platform, or MetaHuman (2023) within Unreal Engine. However, these are less flexible in their design paradigms. AI tools like Deepreel (2023) and D-ID (2023) can generate videos of digital humans: users can create avatars in the platform and animate them with text or spoken word input. The platforms cannot create animated full-body shots of the characters, or rotate them around their axis, though, and the platforms are also not geared towards singing. Overall, the tools are either highly complex or creatively restrictive. To find out more about the options for DIY musicians, a practice-led case study was carried out, as described below.

3. Case Study

To explore the potential of crafting personalized performance avatars through a DIY approach, a case study was undertaken. This study involved the creation of an animated virtual character utilizing 3D graphics software, coupled with AI-powered motion capture techniques. Furthermore, bespoke scenes were created, uniting the artist and the avatar within a shared visual environment. Footage of the artist herself was composited into the scene through a green screening approach. The result was presented in a music video for the song "Bird Caller" (figures 15.1 and 15.2).

FIGURE 15.1   This QR code can be scanned to watch the "Bird Caller" music video.

FIGURE 15.2   A screen shot of the "Bird Caller" music video. The character's movements are controlled by the author's movements.

As an audiovisual artist under the name Nyokee, the author's creative vision is influenced by the aesthetics of video games. Central to the music video's narrative is the Bird Caller character (BC), an anime-style girl with bird wings. BC serves as a metaphor for the artistic calling: her desire to dedicate herself to her art resembles the purposeful navigation of a migratory bird. Visually, the character was created to evoke the aesthetics of collectible vinyl dolls or the visual style of popular Nintendo games.

BC was created within Blender, an open-source 3D computer graphics tool. Next, the character was rigged and animated in Adobe Mixamo (2023) and the AI-driven Plask (2023) motion capture platform. While BC was tested only in the context of a music video, there are potential applications for use on stage, for example by using motion trackers or a motion capture suit. In tandem with the creation of bespoke 3D assets, generative AI was used to create still background images via the Midjourney (2023) platform. The song itself was produced in Logic Pro X. Section 3.1 describes the process of creating and animating BC, and Section 3.2 describes the additional work undertaken. Section 3.3 is a discussion of the creative process within the context of Csikszentmihalyi's flow theory, assessing the challenges and affordances of the DIY approach.

3.1.   Virtual Character Design

Blender (2023) is a free, open-source 3D computer graphics software tool with a broad range of functions ranging from game asset design to video effects. Its expansive capabilities enable artists to model, texture, and animate virtual assets. In the modelling process, artists can use a range of editing tools to sculpt 3D shapes, which consist of points (vertices) in a 3D space. The points are connected with edges to form polygons, forming a mesh (figure 15.3). Next, materials are applied to the mesh, allowing for the definition of attributes like colour, roughness, or transparency. Different objects can be placed together in scenes and illuminated with light sources.

Figures 15.3 and 15.4 show the foundational mesh of BC and its final textured look. The progression into animation required rigging, a process involving the integration of an armature – a bone structure akin to a real-life skeleton – into the virtual character (figure 15.5). By mapping these bones to the mesh, their movements could control corresponding deformations in the mesh. The process of rigging can be complicated. However, the online platform Adobe Mixamo (2023) automates this endeavour in a streamlined way. Mixamo also offers a range of movement presents, some of which were used for dance scenes in the music video (figure 15.6). Plask (2023) is an automatic, AI-powered motion capture tool which the author used to extract dynamic movement data from dance videos (Figure 15.7). This extracted motion was mapped onto the character within Blender, a process illustrated in figure 15.8. In order to control the character live on stage, it could be imported into a games engine, using motion trackers to map real-live movements to the character on-the-fly.

FIGURE 15.3   The underlying mesh of the character.

FIGURE 15.4   The mesh with its final materials applied.

FIGURE 15.5   The final rigged character.

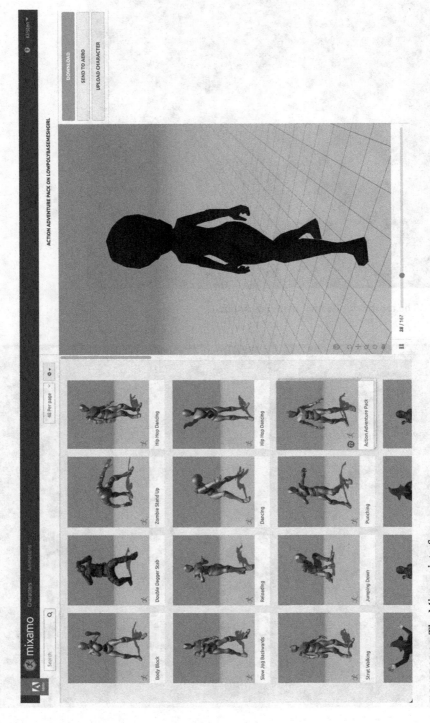

FIGURE 15.6 The Mixamo interface.

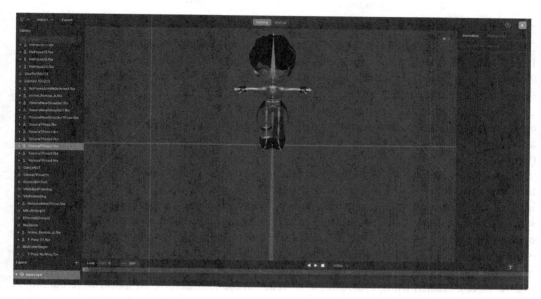

FIGURE 15.7 The Plask interface.

FIGURE 15.8 The character's dance movements are controlled through deformations of the armature, through a mixture of Mixamo presets and Plask motion capture from video.

3.2. *Additional Work*

Continuing her creative journey, the author recorded herself singing and dancing, using an iPhone 11 Pro, set against a blue screen backdrop. In order to insert this new footage into the music video, the blue background needed to be removed. To this end, the author used the AI-powered keying feature within the video editing software Davinci Resolve.

In addition to the character, a small scene was created in Blender, consisting of plants and trees on an isometric floating island, surrounded by clouds and a hot air balloon (figure 15.9). The author also recorded slow-motion videos of birds in flight, which are occasionally shown in the video (figure 15.10).

Midjourney (2023) is a popular generative AI tool built on image diffusion (a type of machine learning) that produces new images according to text prompts entered by the user. By typing "/imagine prompt", a user can request any image, in any style. Further stylistic parameters can be specified, such as image format requirements, or existing images can be used as inputs for new outputs. At the time of writing this chapter, users that are subscribed to the Midjourney platform own the copyright to all images they produce. For this music video, several landscapes were created in the style of Super Mario games (figure 15.11). The author also added abstract art as glitch frames (figure 15.12), created using Procreate on an iPad. In the final stages, all assets were combined within Premiere Pro. In the next section, the creative process is discussed within the context of Csikszentmihalyi's flow theory, to address the question of how accessible the process of creating custom virtual characters is to DIY musicians.

FIGURE 15.9   Floating island in the music video.

FIGURE 15.10   Slow-motion bird footage.

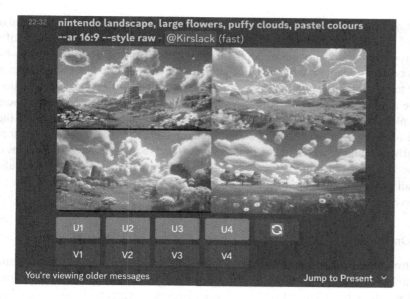

FIGURE 15.11   Background images created in Midjourney through a text prompt.

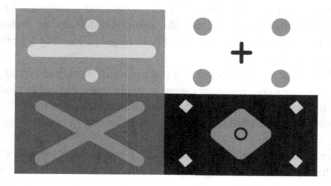

FIGURE 15.12   Four abstract images for quick glitch art edits.

3.3.   *Critical Reflection*

To assess further how accessible the process of creating customs virtual characters is to DIY musicians, this section reflects on the creative process through the lens of Csikszentmihalyi's flow theory (1997, p.107). Csikszentmihalyi is a Hungarian-American psychologist known for his pioneering work in the field of positive psychology. His concept of "flow" refers to a state of heightened focus and immersion in an activity, characterized by the merging of action and awareness. To make a creative journey enjoyable, Csikszentmihalyi stresses that the activities themselves are as important as the manner of their execution. Notably, enjoyable creative endeavours share common attributes. Csikszentmihalyi notes that they are marked either by discovery or by creating something new. The process of designing the character was indeed a source of enjoyment, due to the creative freedom involved, and the discovery of new tools felt empowering. On the other hand, the tools were difficult to learn, which at times hindered flow. In the following sections, Csikszentmihalyi's conditions of flow are used to reflect on the creative process in more detail.

The Clarity of Goals

In order to enter a flow state, the goal of an activity has to be made clear in advance. The author began with a defined concept and reference images, but these were not matched by her skills, and the required processes ended up more time-consuming than anticipated. Therefore, the concept had to be repeatedly adapted, which in turn hindered the flow. Different visual styles in 3D graphics appear to require entirely different technical approaches. Unlike music production, where a track can be remixed and altered easily, any changes to the shape or look of a character might require a complete reworking from scratch. The author created three or four characters that ended up unsuitable for animation, and had to be discarded. Overall, the creative goals and their execution are mutually dependent, which meant that the author had to carry out iterative cycles of conceptualization, and technical planning across several platforms. The inclusion of AI tools also meant that the overall process became more convoluted, and the use of several platforms made it more difficult to arrive at a united visual style.

Knowing How Well One Is Doing

Csikszentmihalyi stresses the importance of immediate feedback during an activity. Within Blender, graphics can be rendered in real time, while modelling, and the creator can see at all times what the final result will look like (figure 15.13). However, because several tools were combined (Blender, Plask, video footage, and so on), the process became fragmented over several devices and platforms, making it more difficult to perceive the entirety of the project in a holistic way. For instance, motion capture was a separate step after filming, so it was not clear until later what kind of footage might work well. Since the author filmed herself, she also could not see the footage while filming. Live motion capture can help with this, but would have required learning a games engine, or expensive hardware trackers.

The author's continuously evolving skill levels also led to a perception of diminished quality. This is similar to music production, where many producers seem to agree that a project is rarely completed, but rather abandoned – the quality of the end product becomes a moving goalpost. Her engagement with online learning materials meant that the author began to second-guess the "right" way of doing things.

In contrast to the collaborative nature of real-life graphic design teams, this project was completed by just one person. This scenario made it difficult to remain unbiased during the

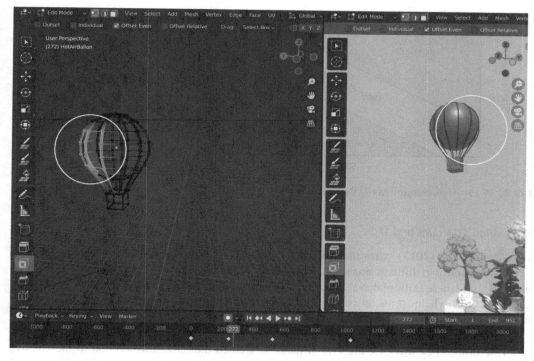

FIGURE 15.13 Modelling while viewing a pre-render in Blender.

FIGURE 15.14 A documentary about the music video.

creative process: for example, the author cannot "unsee" the underlying meshes in the final rendered result. As the author was working alone, feedback from others would have been invaluable. However, this was not possible until after the project was finished, so the feedback was not immediate. This issue could be mitigated to a degree by getting third-party opinions through online forums.

After release, the video led to positive feedback online and an increase in following. Online followers commented on the video, saying that it was "kawaii" and that the visuals were "fun". The author also created a documentary video to explain her creative process (figure 15.14) and viewers commented that there was "a lot of knowledge, well presented" and that they found the author's creative journey of using technology enjoyable.

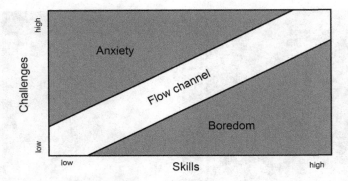

FIGURE 15.15   Csikszentmihalyi's flow channel.

Balancing Challenges and Skills

Csikszentmihalyi (2013) states that what people enjoy is not being in control, but the sense of exercising control in difficult situations. His concept of the "flow channel" (figure 15.15) describes an ideal zone where skills match challenges, fostering immersive engagement. Anxiety arises when challenges overwhelm skills, disrupting this balance and hindering flow attainment. On the other hand, when a task is too easy, boredom can ensue.

The author had been learning Blender intermittently over several years, although not at the requisite proficiency level for the task at hand. It seemed that any new addition to the creative process, particularly when it pertained to animation, required that a large new field of techniques was researched. This is in line with the job specializations within graphics, where specific experts deal with modelling, animation, texturing, rigging, etc., unlike generalist music producers. Csikszentmihalyi stresses that flow states work best when there is no fear of failure, which was difficult to achieve given that the activity was in many ways new and many skills had to be acquired on-the-fly – this disrupted the creative process. Overall, the author found herself in the flow channel where the required workflows were familiar and predictable, but less so when problem-solving was required, which was often the case.

The Merging of Action and Awareness

Csikszentmihalyi explains that flow states are characterized by a merging of action and awareness. Individuals will experience a state of intense focus and engagement and become fully absorbed in an activity, experiencing a sense of effortless control. This indeed happened where the author was able to align project sub-goals with well-defined workflows and tools she understood. However, when learning new aspects of the tools and troubleshooting challenges, the author encountered what Csikszentmihalyi terms "psychic entropy", an unpleasant state of mental turbulence or disorder.

Avoiding Distractions

With the abundance of available tools, platforms, and creative avenues, it was difficult to avoid distractions. The author had wanted to experiment more with graphics for many years, and the weight of many unrealized visions, in contrast to the difficulty of the process contributed to a sense of overwhelm. Due to the creative process being time-intensive and technically complex, it was sometimes difficult to maintain focussed attention for the necessary duration. However, setting a tight deadline, and a sole focus on the task helped the author remain focussed, and this proved very useful.

Forgetting Self, Time, and Surroundings

Due to the high pressure of the deadline, the author did experience flow states in the project. She ended up working for several very long days (up to 16 hours) to complete her vision, and when the project went to plan, she felt herself to be very immersed in the process.

Creativity as autotelic experience

According to Csikszentmihalyi, an autotelic experience refers to a state of intrinsic motivation and deep engagement in an activity for the sheer enjoyment and satisfaction it provides, rather than for any external rewards or outcomes. In such an experience, the activity itself becomes its own reward, and the individual is fully absorbed and immersed in the task at hand. The term "autotelic" is derived from the Greek words "auto" (self) and "telos" (goal), reflecting the self-contained and self-rewarding nature of the experience. Despite the challenges, the author continues to feel inspired to create visual assets for music videos and live performances. This is because she feels an intrinsic sense of fulfilment from the activities.

4. Discussion

Overall, it is possible for musicians to create 3D avatars using a DIY approach. There is an abundance of affordable tools and online learning materials. As a result, artists can create unique visual identities and enjoy a new kind of creative freedom in the visual domain, which can also form part of impactful marketing strategies.

Musicians will, however, face challenges in trying to map their familiar creative processes to the workflows required for 3D graphics. Holistic 3D graphics tools and games engines are complicated and geared towards teams of specialists, while DAWs allow creatives to create acceptable outputs alone, and more quickly. In the domain of graphics, it is more difficult to alter a creative vision during its technical execution, as different artistic styles can require entirely different technical approaches. Third-party character design tools are easier to operate. However, these platforms tend to have less flexibility in their visual styles: VRoid Studio is for anime characters, MetaHuman is geared towards realistic humans, and Ready Player Me produces human characters reminiscent of specific video games. If motion capture from video is required, another, separate tool needs to be used. In the context of animation, Mixamo is highly useful for quickly rigging and animating a character, but if the rig is not quite right, or if a movement preset needs to be tweaked (such as "walk an extra step to the left"), an artist must embark on the difficult and time-intensive journey of learning rigging and animation from scratch, or invest in expensive hardware. Similarly, most AI tools will only help with one building block (for example motion capture, but not character modelling). In contrast, most DAWs can be used to produce music in almost any genre, from start to finish. When working with music samples, or plugins, it is easy to cut or slice segments, play them at once, or combine them in other ways: music production tools and techniques allow for cross-compatibility, while graphics tools can be hard to combine. Graphics tools are also geared mainly towards video games, and towards film and animation, and less so towards music videos or musical live performances – at least at the time of writing this chapter.

This paper is only an account of one person's experience, and more case studies are required to fully understand the overall situation. While the author recognizes that her many years of musical training will make music production appear easier than 3D graphics, much of the above discussion is reflected in online forums and discussions with other creatives. All this underscores the necessity for enhanced and more user-friendly tools, such as accessible character-creation tools specifically for musicians. New market developments will likely help bridge this gap, such as Sony's Mocopi (2023) or Move AI (2023), both of which will simplify motion capture at home.

5. Conclusion

Virtual avatars are a powerful means to creating unique artist personas. They work well as marketing tools, and empower musicians to take control of their creative identities. The purposeful artifice of virtual characters can be particularly empowering for artists, as it caters to the widespread fascination with role playing, sentient machines and discussions around hybrid creativity. New 3D graphics tools, AI and even video games offer a vast portfolio of possibilities for musicians to create their own virtual characters for free. The options include: game engines, 3D graphics tools, third-party character creators, social media apps, video games, chat platforms and VR applications.

However, these tools can be creatively restrictive, or difficult to learn, which creates challenges in the creative process. While many creative issues can be resolved by combining several tools, this leads to switch tasking and issues in cross-compatibility. Overall, musicians must think critically about their creative process and find their own patch in a plethora of possibilities. There appears to be a market gap for more powerful, and more user-friendly avatar creation tools that are specifically geared towards musicians. On the other hand, while creating virtual avatars is not a skill that is equally accessible to everyone, those who do embark on this journey are rewarded with the ability to forge a unique and remarkable visual style.

References

ABBA Voyage (2022). (website), available online from https://abbavoyage.com/ [accessed August 2023].
Acid Horizon (2021). PC Music, Accelerationism, and Xenofeminism (online podcast), available online from www.youtube.com/watch?v=EtKODfEG58U&t=7s [accessed August 2023].
Ajeesh, A. K. and Rukmini, S. (2023). 'Posthuman perception of artificial intelligence in science fiction: an exploration of Kazuo Ishiguro's Klara and the Sun', *AI and Society*, Vol. 38 (2), pp. 853–860.
Auslander, P. (2009). Musical Persona: The Physical Performance of Popular Music. In: Scott, D. B. (Ed.), *The Ashgate Research Companion to Popular Musicology*, New York, Routledge, pp. 303–316.
Blender (2023). (website), available online from www.blender.org/ [accessed August 2023].
Born, G. and Hesmondhalgh, D. (2000). Introduction: On Difference, Representation, and Appropriation in Music. In: Born, G. and Hesmondhalgh, D. (Eds.), *Western Music and Its Others: Difference, Representation, and Appropriation in Music*, University of California Press, pp. 1–58.
Carroll, N. (2021). (website). 'Augmented intelligence: an actor-network theory perspective', *ECIS 2021 Research Papers*, Vol. 37, available online from https://aisel.aisnet.org/ecis2021_rp/37 [accessed August 2023].
Crypton (2023). (website). Who is Hatsune Miku?, available online from https://ec.crypton.co.jp/pages/prod/virtualsinger/cv01_us [accessed August 2023].
Csikszentmihalyi, M. (1997). *Creativity: Flow and the Psychology of Discovery and Invention*, New York, Harper Perennial.
Csikszentmihalyi, M. (2013). *Flow: The Psychology of Happiness*, New York, Penguin Random House.
D-ID (2023). (website), available online from www.d-id.com/ [accessed August 2023].
Dautenhahn, K. (2007). A Paradigm Shift in Artificial Intelligence: Why Social Intelligence Matters in the Design and Development of Robots with Human-Like Intelligence. In: Lungarella, M., Iida, F., Bongard, J., Pfeifer, R. (Eds.), *50 Years of Artificial Intelligence*, Springer, Berlin, Heidelberg, pp. 288–302.
Deepreel (2023). (website), available online from www.deepreel.com/ [accessed August 2023].

Deleuze, G. and Guattari, F. (1988). *A Thousand Plateaus: Capitalism and Schizophrenia*, UK, Bloomsbury.

Diederichsen, D. (2015). Sound and Image Worlds in Pop Music. In: Daniels, D. and Naumann, S. (Eds.), *Audiovisuology, A Reader*, Vol. 1: Compendium, Vol. 2: Essays, Verlag Walther König, Köln., pp. 554–581.

Edlom, J. (2020). Authenticity and Digital Popular Music Brands. In: Tofalvy, T. and Barna, E. (Eds.), *Popular Music, Technology, and the Changing Media Ecosystem*, London, Palgrave Macmillan, pp.129–145.

Facetune (2023). (website), available online from www.facetuneapp.com/ [accessed August 2023].

Frith, S. (2001). Pop Music. In: Frith, S., Straw, W. and Street, J. (Eds.), *The Cambridge Companion to Pop and Rock*, Cambridge, Cambridge University Press, pp. 93–108.

Gioti, A. (2020). 'From artificial to extended intelligence in music composition', *Organised Sound*, Vol. 25 (1), pp. 25–32.

Gladwell, M. (2002). *The Tipping Point: How Little Things Can Make A Big Difference*, New London, Abacus.

Godin, S. (2005). *Purple Cow: Transform Your Business by Being Remarkable*, UK, Penguin.

Haraway, D. J. (1991). A Cyborg Manifesto: Science, technology, and Socialist-Feminism in the Late Twentieth Century. In: *Simians, Cyborgs, and Women: The Reinvention of Nature*, Simians (ed.) New York, Routledge, pp. 149–181.

Hermes, K. (2022). *Performing Electronic Music Live*, New York, Routledge.

Hertzmann, A. (2018). 'Can computers create art?', *Arts*, Vol. 7 (2), p. 18.

Holly+ (2023). (website), available online from https://holly.plus/ [accessed August 2023].

Horizon Worlds (2023). (website), available online from www.meta.com/gb/horizon-worlds/ [accessed August 2023].

Huttu-Hiltunen, I. (2020). *Evolution of Animated Characters in League of Legends World Championship Opening Ceremonies: Which Technologies Are Used to Impress the Audience, and How Does the Use of Technology Evolve between 2017 to 2019?*, Online master's thesis, available online from www.theseus.fi/handle/10024/343430 [accessed March 2023].

IMVU (2023). (website), available online from https://secure.imvu.com/welcome/ftux/ [accessed August 2023].

Lord, A. (2018). (website). Why AI pop stars are even more real than the real thing. *Quietus*, available online from https://thequietus.com/articles/24629-lil-miquela-poppy-ai-pop-stars [accessed August 2023].

Magnopus (2021). (website). Madison Beer: immersive reality concert, available online from www.magnopus.com/projects/madison-beer [accessed August 2023].

Mántaras (2020). Artificial Intelligence and the Arts: Toward Computational Creativity. In: de Grey, A. and Rossiter J. (Eds.), *The Next Step: Exponential Life*, USA, BBVA-Open Mind.

Middleton, R. (1990). *Studying Popular Music*, UK, Open University Press.

Midjourney (2023). (website), available online from www.midjourney.com/ [accessed August 2023].

Mixamo (2023). (website), available online from www.mixamo.com/ [accessed August 2023].

Mocopi (2023). (website), available online from https://electronics.sony.com/more/mocopi/all-mocopi/p/qmss1-uscx [accessed August 2023].

Morris, M. N. (1971), *Writings on the General Theory of Signs*, Den Haag, Mouton.

Move AI (2023). (website), available online from www.move.ai/ [accessed August 2023].

Peterson, R. (1997). *Creating Country Music, Fabricating Authenticity*, Chicago, University of Chicago Press.

Plaete, J., Bradley, D., Warner, P. and Zwartouw, A. (2022). 'ABBA voyage: high volume facial likeness and performance pipeline', *Siggraph*, Vol. 2022, No. 18. pp. 1–2.

Plask (2023). (website), available online from https://plask.ai/ [accessed August 2023].

Ready Player Me (2023). (website), available online from https://readyplayer.me/ [accessed August 2023].

Rovinelli, J. D. (2021). (website). Sophie's triumphantly plastic music moulded a new world for trans people. *The Guardian*, available online from www.theguardian.com/music/2021/feb/02/sophie-triumphantly-plastic-music-moulded-a-new-world-for-trans-people [accessed August 2023].

Susskind, D. (2020). *A World Without Work: Technology, Automation and How We Should Respond*, London, Allen Lane.

TikTok (2023). (website), available online from www.tiktok.com/ [accessed August 2023].

Unity (2023). (website), available online from https://unity.com/ [accessed August 2023].

Unreal Engine (2023). (website), available online from www.unrealengine.com/ [accessed August 2023].

VRChat (2023). (website), available online from https://hello.vrchat.com/ [accessed August 2023].

VRoid Studio (2023). (website), available online from https://vroid.com/en/studio [accessed August 2023].

VTuber Maker (2023). (website), available online from https://store.steampowered.com/app/1368950/VTuber_Maker/ [accessed August 2023].

Wingström, R., Hautala, J. and Lundman, R. (2022). 'Redefining creativity in the era of AI? Perspectives of computer scientists and new media artists', Creativity Research Journal, 36(2), 177–193. https://doi.org/10.1080/10400419.2022.2107850.

16

TRANSLATING ARTWORKS INTO MUSIC

Synaesthetic Reverse-engineering in Music Composition

Corin Anderson

1. Introduction

Synaesthesia is a perceptual phenomenon experienced by around 4.4% of the population (Simner et al., 2006) where the stimulation of one sensory modality elicits a response in another (Asher and Carmichael, 2013). Auditory-visual synaesthesia is one type of the phenomenon, where the perception of sound induces a visual, as well as auditory, experience for the affected individual. As an auditory-visual synaesthete myself, all sounds automatically and involuntarily induce the visual perception of coloured and textured three-dimensional shapes in my mind's eye.

Synaesthetic visual percepts are often called photisms (Colman, 2008; Cytowic, 2018). I find that musical sounds induce particularly vibrant, definable photisms. During any musically-triggered synaesthetic experience, each of the photisms I visualise correspond to a separate perceived sound. Thus, when listening to a typical four-piece rock band, for example, I will simultaneously 'see' separate photisms for the vocal, guitar, bass, and each drum and cymbal. The appearance of each photism in my mind's eye is determined by various musical parameters, as shown in Table 16.1.

Evidently, sound timbre is the musical parameter with the biggest impact on the appearance of my photisms. Due to the infinite number of timbres that can be synthesised in the making of electronic music, there is no upper limit on the number of synaesthetic visual possibilities that might be induced. I compose and produce electronic music, perhaps in part due to this very reason; the photisms induced by the constantly evolving timbres of electronic music make for a profoundly dynamic and unpredictable visual accompaniment to the music.

I wanted to discover how my auditory-visual synaesthesia affected my music composition processes and outputs, and so conducted autoethnographic research into my creative practice. I undertook three compositional and autoethnographic studies to ascertain whether it was possible to translate music into visual artworks and, by reverse-engineering my synaesthesia, visual images into sound palettes to use as the building blocks for new compositions. Retrospectively, I have chosen to classify these studies as 'descriptive', 'analytic', and 'aesthetic', due to their purpose.

In my 'descriptive' study, I intended to show others what it was like to experience auditory-visual synaesthesia, by creating animations that represented my synaesthetic visualisations of

DOI: 10.4324/9781003396550-16

TABLE 16.1 My auditory-visual synaesthetic correspondences.

Auditory inducer	Visual concurrent(s)
Timbre of sound	Shape, colour, texture, state of matter, size, and weight of photism.
Pitch of sound	Brightness, size, weight, and position of photism on Y axis
Loudness of sound	Size, weight, and position of photism on Z axis
Stereo position of sound	Position of photism on X axis
Reverberation of sound	Shadow and position of photism on Z axis
Duration of sound	Duration of photism
Rhythm of sound	Rhythm of flickering, flashing, and pulsation of photism
Musical note of sound (if known)	Colour hue of photism
Key of music (if known)	Colour hue of all photisms

two of my compositions. The study is 'descriptive' since its aim was to visually *describe* my synaesthetic experiences by translating from the sonic realm to the visual.

The purpose of the 'analytic' study which followed was to explore whether I could work backwards and convert an entire visual artwork into a work of music. It is described as 'analytic' since it aimed to *analyse* my attempts to translate entire visual stimuli into musical compositions.

With my later 'aesthetic' study, I translated only specific aspects of visual stimuli into musical ideas, selectively choosing from these only (what I considered to be) the best ideas to form the bases of new compositions. It is described as 'aesthetic' since its aim was to create compositions that were *aesthetically* pleasing to me, by selecting and translating only certain elements of visual stimuli into musical ideas.

2.   Background and Related Work

To date, there have been several retrospective discussions on synaesthetic artists' work (e.g.: Bernard, 1986; Harrison and Baron-Cohen, 1994; van Campen, 1999, 2010), some first-person accounts of the impact synaesthesia has on individuals' perceptions of music (e.g.: Day, 2013; Rudenko and de Córdoba Serrano, 2017; Vanhalen, 2020) and a small number of academic first-person accounts by synaesthetic artists on how their experiences influence and affect their creative practice (e.g.: Lee, 2018; Püschel, 2017). Several contemporary musicians have identified themselves as synaesthetes and discussed the influences their experiences have on their creative processes. Pharrell Williams claims to have been inspired by his synaesthetic experiences when writing lyrics (Seaberg, 2011), while Lorde visualises her ideas for new songs in her mind's eye and describes the songwriting process as "getting the actual thing to sound like what I've been seeing" (as quoted in Weiner, 2017). However, aside from a few anecdotal accounts such as those mentioned above, there has to date been no scholarly research conducted into how synaesthesia can affect musicians' composition processes and outputs. By providing an insight into how my synaesthetic experiences can affect my composition processes and outputs, this work makes an original contribution to knowledge in a previously unexplored intersection between (electronic) music composition practice, synaesthesia, and autoethnography.

3.   Methodology

This work is a hybrid of artistic research and autoethnography. The knowledge I am interested in attaining is practical knowledge; "learning through doing", or "know how" (Nelson, 2013, p. 42). Nelson (ibid., p. 9) describes this type of knowledge as that "which might primarily be demonstrated in practice – that is, knowledge which is a matter of doing rather than abstractly conceived and thus able to be articulated by way of a traditional thesis in words alone". My practice is in itself a research method and, through this research, I seek to understand how my thinking affects my doing (ibid.); how my synaesthetic perceptions affect my creative practice.

At the centre of this research is an interpretivist philosophy: a research paradigm or "a lens through which we view the world" (Collins, 2010, p. 38). According to Collins, interpretivism is concerned with meaning-making, "reject[s] the objectivist view that meaning resides within the world independently of consciousness" (ibid., p. 38), and attests that meaning resides "between a conscious, meaning-making subject and the objects that present themselves to our perception" (ibid., p. 39). I intend to make (musical) meaning from my synaesthetic experiences: my own unique perceptions of the world around me. I acknowledge my subjective viewpoint as both researcher and subject, but consider the adoption of both roles an unavoidable inevitability due to my unique perceptions of the world; no two synaesthetes' experiences are identical and so any attempt to generalise my findings would be misguided.

4. Methods

During and shortly after each study, systematic self-observational data on my synaesthetic experiences and composition processes were gathered using autoethnographic recording methods proposed by Chang (2008). I analysed and interpreted the data by searching for mentions of translative processes I had utilised and reflective, evaluative thoughts I had recorded. I also compared my descriptions of my own translative processes with those employed by other synaesthetic artists and framed my findings within existing theories.

5. Descriptive Study

For my initial study, I attempted to translate one of my music compositions into a video animation. My intentions with this animation were: to convey a sense of what it is like to experience auditory-visual synaesthesia while listening to music, for the benefit of those who did not experience the phenomenon; and to gain a better understanding of the crossmodal correspondences of my synaesthesia.

My first intention was informed by a study by Ward et. al (2008), which found that animations created to accompany music by synaesthetic artists were preferred over animations by non-synaesthetes. Computer animation technologies enable synaesthetic artists like Carol Steen, David Hockney, and Marcia Smilack to share a visual representation of their synaesthetic experiences with others (Steen and Berman, 2013). By producing animations that represented my synaesthetic visualisations of music, I had hoped that I would be able to give others the opportunity to experience what it is like to have auditory-visual synaesthesia, in part to enhance their enjoyment of the music and also to educate them about the phenomenon.

My second intention, meanwhile, was inspired by an autoethnographic data-generation exercise proposed by Chang (2008). For this, Chang encourages the researcher to draw a "place of significance that helped [them] gain an understanding of [themselves]" (p. 87) before identifying objects in the drawing and explaining their significance. I adapted this exercise by drawing and animating my synaesthetic experiences of the music. By identifying and explaining each individual photism present in the animation, I hypothesised that I would gain a better understanding of the mappings between my auditory and synaesthetic visual perceptions. I composed a piece of music, *Glue* (see Figure 16.1), with the intention to animate it thereafter.

FIGURE 16.1 QR code linking to *Glue* composition.

Using a Huion H420 graphics drawing tablet and Krita, a 2D animation program, I created a simple frame-by-frame animation to represent the lead synthesiser in the track. Using my digital audio workstation (DAW) to play back the music, I looped the first bar, and drew the resulting synaesthetic photism and its motion, choosing the pen style, thickness, and colour that were best able to reproduce the photism's appearance. I repeated this process for each bar and made notes throughout explaining my auditory-visual translations.

5.1. Findings and Discussion

I found that translating my synaesthetic experiences of music into computer-generated animations resulted in the experience of 'Martian colours' (Ramachandran and Hubbard, 2001): colours that "are somehow 'weird' or 'alien' and don't look quite the same as normal 'real world' colours" (p. 26). I would describe this experience as something similar to the uncanny valley effect (Mori, 2012): a term often used to describe the uncomfortable response in the observer upon perceiving an object (real or virtual) which closely, but not quite perfectly, resembles a human being. No matter how hard I tried to match my visual creation to the photism, its movement, colour, and surface texture, while very similar to those of the synthesiser's corresponding photism, ultimately fell short of accurately portraying the synaesthetic experience I had when listening to the music. This uncanny valley effect could at least partially explain why this animation failed to adequately represent my synaesthetic visualisations.

My inability to accurately recreate my synaesthetic photisms in animation software may simply be due to my lack of skill and experience with computer animation. However, it is not uncommon for synaesthetes to report 'Martian colours' (Ramachandran and Hubbard, 2001) that do not quite match 'real-world' colours. For example, Püschel (2017) experienced this effect when attempting to pick from a limited colour palette the exact colours she synaesthetically perceived when emotionally engaging with certain photographs. This is exactly how I felt when I saw that the animation produced for *Glue* was not exactly how I visualised the music.

While creating the animation, I found myself asking whether there was much point in painstakingly recording my synaesthetic experiences in the form of computer-generated animation. Cavallaro postulates that

> just as no two synesthesias are ever quite identical despite possible similarities in their manifestations, so no two intuitions of the dynamic properties of sound will ever be likely to yield identical visualizations. This is clearly borne out by the impressive diversity found in visual works by synesthetic artists inclined to give shape to sound.
>
> *(Cavallaro, 2013, p. 96)*

If synaesthetic visualisations are so personal and individual, who am I to dictate what others should 'see' when listening to my music? I, for one, have found animations created by synaesthetes to accompany other artists' music to be jarring, since their interpretations of the music do not resemble my own. I would therefore imagine that other people (synaesthetic or not) may find any animations created or directed by me to diverge from their own ideas of what the music should 'look' like.

Furthermore, in trying to be as accurate as possible in my translating from music to visuals, and 'stay true' to my synaesthetic experiences, the pleasure I normally feel while musicking (i.e., when composing, producing, and listening to music) was absent. I found that there was little 'creation' in what I was doing. Abbado (2017, p. 136) suggests that "specific correspondences [between auditory and visual mediums] can also act as a form of cage, providing constraints that are sometimes uninspiring". In creating these animations, it felt like I was artlessly and clinically translating my compositions from one medium to another. I concluded that now was not the time to start learning how to animate and to instead focus on finding a way to use my synaesthetic experiences as inspiration for new musical compositions. I am, after all, a composer of music and not an animator. I was keen to discover whether I could exploit my synaesthetic experiences to inspire the creation of new music.

6. Analytic Study

By translating in the opposite direction from visual to sonic, I had to 'reverse-engineer' my auditory-visual synaesthetic experiences. This was achieved by deconstructing the painting into identifiable shapes, then audiating sounds that induced photisms resembling these shapes, before producing and recording these sounds in my DAW.

Most synaesthetes experience synaesthesia in a single direction (Teichmann et al., 2017; Steen and Berman, 2013). I automatically and involuntarily visualise music as a landscape of various shapes with colour and texture, but cannot hear music when looking at a painting. However, over the years, I have come across many visual artworks that resemble my auditory-visual synaesthetic experiences and have recently inspired the composing of new works of music, such as *Composition A XXI* by László Moholy-Nagy (1925). It is this style of abstract, geometric art that I tend to be drawn to, due to its similarities to the scenes that I 'see' in my mind's eye whenever I listen to, compose, and perform music. I may not be able to accurately translate my synaesthetic experiences into visual artworks but observing these artworks which exist in the 'real world', unlike my synaesthetic visualisations, somehow seems to validate my experiences, confirming that they are not merely a figment of my imagination.

Following on from my attempts to translate music to a visual artform, I wanted to see if it was possible to work in the opposite direction and translate visual artworks that resembled my synaesthetic experiences into new music compositions, for the purpose of better understanding my synaesthesia and the role crossmodal translation has in my compositional processes.

I listed all of the auditory-visual mappings in my synaesthesia (see Table 16.1). Using this as a translation guide, I attempted to convert László Moholy-Nagy's (1925) painting Composition *A XXI* into music, due to its strong resemblance to the synaesthetic experiences I have while listening to music. I began by mentally separating the painting into identifiable visual objects (shapes). For each visual object, I audiated (heard with my mind's ear and gave musical meaning to (Gordon, 1999)) a range of sounds until I discovered one which induced a synaesthetic photism resembling the visual object. Using my synthesisers and DAW, I then produced and recorded this sound. The pitch of the sound was determined by the approximate vertical position of the shape on the canvas, while the sound's placement in the stereo field was determined by the shape's

FIGURE 16.2 QR code linking to *Composition A XXI* (Moholy-Nagy, 1925) animation.

approximate horizontal position. In order to translate the still image into music (a time-based medium), I injected motion into my perception of the painting by imagining the different shapes moving in a sequenced loop around the canvas. Steen and Berman (2013) discuss how synesthetic artists who translate moving photisms (such as those induced by listening to music) to static visual artworks (such as paintings) make use of a variety of methods to convey a sense of movement in their work. These include morphing, sequencing, and layering shapes. While these artists face the challenge of translating a fluid, moving medium into a fixed, static one, I wanted to do the opposite: translate a still image into an evolving work of music. Since I was keen to ensure that the resulting synaesthetic experience of the music I had composed closely resembled the painting I was translating into music, any movement that I added was intentionally minimal and repetitive through use of looping. I created a simple animation that demonstrated this movement and later synchronised the music I had composed to it (see Figure 16.2).

6.1. Findings and Discussion

The music I composed was intended to be a fairly precise translation of the *Composition A XXI* (Moholy-Nagy, 1925) painting into music. Listening back to the music causes me to visualise all the shapes present in the artwork, and so I would argue that the study was, by this measurement, successful to a great extent. The main difference between my synaesthetic visualisation of the music I composed and the painting is that the photisms move in loops around the canvas, while the shapes in the painting are fixed in place. However, the insertion of perceived motion into the painting was the only time I felt I had engaged in any sort of creative decision-making during the compositional process. Interpreting colourful shapes on the canvas as timbres with pitch and stereo position felt procedural, akin to translating a sentence from one language to another. Furthermore, the resulting music composition was extremely repetitive due to the limited movement I had allowed myself to introduce to my audiations.

Although I was displeased with the music I had produced and the lack of creativity involved in the composition process due to the limits I had set myself, I did, however, discover that it was possible to 'reverse-engineer' (to borrow a term from the engineering discipline meaning to disassemble in order to analyse) my synaesthetic experiences through a process of audiation.

> Audiation takes place when we hear and understand in our minds music that we have just heard performed or have heard performed sometime in the past.. . . We also audiate when we hear and understand in our minds music that we may or may not have heard but are reading in notation or are composing or improvising.
>
> *(Gordon, 1999, p. 42)*

Hubbard (2010) opines that Beethoven – who, at this point, had become deaf – likely made use of auditory and musical imagery and notational audiation while composing his Ninth Symphony, "to create or simulate auditory qualities" of the music (p. 316). I would argue that audiation is crucial to the process of reverse-engineering my synaesthesia. When attempting to translate a visual image into sound, I ask myself 'what sound would induce a synaesthetic photism that resembles this shape?', then audiate a range of sounds until I find one that synaesthetically induces a photism resembling the shape. In my mind's ear, again through a process of audiation, I then 'fine-tune' this sound, so that its photism more closely matches the shape. I can then produce this sound and record it in a DAW.

To summarise, my synaesthesia gives me the ability to visualise audiations of music. With a specific visual image in mind (e.g., a painting), I can audiate a range of sounds and select one whose corresponding synaesthetic photism resembles the visual image. This enables me to 'reverse-engineer' my auditory-visual synaesthesia, allowing me to translate a visual image into a musical idea, which can then be produced using a musical instrument such as a synthesiser and recorded in DAW.

There is a significant lack of research into synaesthetic reverse-engineering for artistic purposes. One artist who has engaged in synaesthetic reverse-engineering is lexical-gustatory synaesthete James Wannerton (he tastes flavours upon reading or hearing words). On one occasion, Wannerton reverse-engineered his experiences in order to translate the flavours of ingredients for various meals into objects to be photographed for food journal *The Gourmand* (Nowness, 2013). Wannerton explains:

> Creating images of tastes is quite difficult because it requires a bit of reverse engineering and that for me is the clever part. As my synaesthesia, like most others, runs one way only, in order to create an accurate image I have to eat a particular food (the subject of the image) and read pages of text until a corresponding word "pops up" that produces a taste and texture that matches the food I'm eating.
>
> *(Wannerton, 2015, p. 30)*

My attempts at reverse-engineering are similar to Wannerton's, in that I work backwards from the concurrent to the inducer, asking myself questions like: 'What sound looks like a red circle?'.

7. Aesthetic Study

With the two compositional studies I have so far discussed in this paper, my intention was to, as accurately as possible, translate an entire creative work from the medium of music into visual art, or vice versa. Since the creative work had already been produced, translating it from one medium to another felt to me like nothing more than a procedural exercise. I felt little joy in the translation process and was not particularly proud of the creative outputs I produced, due to the limited role my artistic input played in the process. According to Steen and Berman (2013, p. 673), "the synesthetic artist must filter [their] experiences to generate the final artistic product". They go on to argue that the synaesthetic artist "must make informed decisions about the usage or portrayal of [their] visions, and this requires a type of self-observation" (p. 673). When creating the following compositions this section discusses, I chose to filter my synaesthetic experiences, only translating specific aspects of visual images into musical ideas, to allow for more artistic input and creative freedom in the composition process.

The composition process for my works *A XXI*, *Shiraga*, and *Infinity Pyramids* involved the use of selective translation, based on my aesthetic preferences, most prominently in the initial sound design stage. To help craft sound palettes and inform the musical direction I would take each composition in, visual images were translated into musical ideas through a process of synaesthetic reverse-engineering involving audiation (as explained in the previous section). These musical ideas formed the bases of each composition and are discussed below.

7.1. Findings and Discussion

Dissatisfied with my initial attempt at translating *Composition A XXI* (Moholy-Nagy, 1925) into music, I experimented with rearranging and remixing some of the sounds from this early draft into a coherent composition that more closely aligned with my own musical tastes and compositional style, which I named *A XXI* (see Figure 16.3).

I was much more satisfied with this composition. This led me to the realisation that it did not ultimately matter whether my synaesthetic visualisation of the finished music work resembles the visual artwork that inspired it. I realised that, by choosing only certain elements of a visual stimulus to translate into a sonic palette for use in a musical work, the resulting freedom afforded me more creative input in the compositional process. With my previous attempt at translating a visual artwork into a work of music – my 'analytic' audiovisual interpretation of *Composition A XXI* (Moholy-Nagy, 1925) – I had made an effort to translate *all aspects* of the painting into musical ideas, limiting the artistic freedom I had in the compositional process. The only creative decisions that I can identify in the production of this work were the injection of motion into my audiations and the temporary muting of various parts in the mix. Otherwise, the work of music I produced was, for me, an accurate synaesthetic translation of the entire painting into music.

On the other hand, only *some* of the musical ideas present in *A XXI*, *Shiraga*, and *Infinity Pyramids* were translated from visual stimuli. These compositions were not intended to be precise translations of *entire* visual artworks. They were instead created for aesthetic purposes, with translation between the visual and aural modalities only coming *at the start* of the compositional process as a means to generate initial musical ideas to be further developed into full works of music. The underpinning musical ideas that formed the bases of these compositions were translated from *specific* aspects of visual stimuli, by audiating sounds that synaesthetically induced photisms resembling the visual stimuli and producing and recording those that I believed had the most musical potential.

An example of this approach can be found in the creation of *Shiraga*. This composition stemmed from a 'found sound' recording of a beeping electronic door lock. This sound induced a synaesthetic photism resembling a metallic red disk. I searched for artworks resembling a metallic red disk and discovered Kazuo Shiraga's *Sorin* (1970). Captivated by the forms, colours, and

FIGURE 16.3   QR code linking to *A XXI* composition.

textures used in the artwork, I searched for further examples of Shiraga's work. I found several other paintings that interested me and used these and *Sorin* as inspiration for the palette of sounds I would use in my composition. I reverse-engineered my auditory-visual synaesthesia and worked backwards, audiating the sounds that would generate photisms similar to some of the forms in the paintings. I then produced and recorded these sounds in my DAW and composed short loops with them. To produce a synaesthetic visual experience that resembled the 'smeared paint' appearance of Shiraga's work, I ran each loop through various combinations of distortion, reverb, delay, and flanger effects (usually with the feedback turned up high). This provided me with the sound palette that I then used to construct this composition. In this instance, the forms, colours, and textures in the paintings were translated into several timbres that were then used to construct this *Shiraga* composition (see Figure 16.4).

My synaesthetic perception of the completed composition resembles, to a certain extent, the Shiraga paintings it was inspired by. The 'beeping' found sound – the focal point of my composition – induces a photism similar to a metallic red disk, much like the centrepiece of Shiraga's *Sorin* (1970). The synthesisers that complement the found sound bring to mind red, orange, and yellow splodges, with a texture of smeared paint, much like the colourful streaks in the painting. The percussive elements in the composition (the electronic drums and cymbals) are the exception, as the photisms they induce (brown wooden blocks and silver blades) do not resemble any aspect of the painting. I do not think that it matters that my synaesthetic experience of the finished composition does not ultimately look *exactly* the same as the Shiraga artworks I initially intended to base the composition on. My decision to add electronic drums and percussive elements, the synaesthetic photisms of which diverge drastically from the appearance of the paintings, was likely due to my own musical biases towards rhythmically-driven electronic music. As a producer and fan of electronic dance music, I must not ignore, dismiss, or be ashamed of the undeniable influence of my own production and listening preferences. Indeed, there are many similarities between my latest compositions and music by Jon Hopkins (2018) (e.g., *Singularity*), Nathan Fake (2020) (e.g., *Blizzards*), Rival Consoles (2018) (e.g., *Persona*), and Bicep (2021) (e.g., *Isles*) that I have undoubtedly been inspired by, and I believe that it is important to acknowledge these external influences.

In another case, I found that it was possible to translate in both directions, from visual to aural *and* aural to visual, to provide further inspiration in the composition process. For my composition *Infinity Pyramids*, I was interested in composing a piece of music using an original visual stimulus as inspiration (as opposed to translating elements of a pre-existing visual stimulus such as a painting). I found an infinity symbol (∞) I had recently sketched in a notepad and decided to translate this into a sound to be used in a new composition. I followed my typical visual-to-audio translation

FIGURE 16.4   QR code linking to *Shiraga* composition.

FIGURE 16.5 QR code linking to *Infinity Pyramids* composition.

process by audiating a sound that synaesthetically induced a photism resembling the sketch, before synthesising this sound in my DAW. Not quite content with the sound this translated into, I made some modifications to the synthesiser patch and reflected the corresponding synaesthetic visual changes in my notepad by amending the sketch. Keen to explore this back-and-forth approach further, I made further alterations to the sketch and mirrored these changes in the synthesiser patch. This process continued until I was happy with the synth sound that became the catalyst for my composition *Infinity Pyramids* (see Figure 16.5).

Reflecting on the creation of the compositions discussed above, it is clear that my synaesthetic experiences play a key role primarily in the initial sound design stage of the composition process, the purpose of which is to provide me with a palette of sounds to work with in the arrangement stage that follows. This is because, with the compositions discussed above, I made the conscious decision to 'tap into' my synaesthesia and translate images to sounds and vice versa to inspire the creation of new sounds. However, it is difficult to determine whether my synaesthetic visualisations of the music affect the decisions I make when arranging the sounds and musical ideas I have produced into a coherent structure, in the arrangement stage that follows. Attempting to interrogate my creative process interrupts my state of flow and so it is difficult to gather data on whether my synaesthetic experiences influence my decision-making during the arrangement stage of the composition process. I would suggest that, in this arrangement stage, my musical decision-making is at least partly informed by which combinations and configurations of shapes, colours, and textures 'look' the most appealing, *in addition to* which combinations and configurations of timbres, rhythms, and melodies I perceive to be the most enjoyable to listen to. However, I should note that I do not consciously separate the aural and (synaesthetic) visual experiences of the music into two separate sensory streams when making these creative decisions. The music is instead perceived as one auditory-visual 'whole' – a gestalt – and it is this overall experience that informs my decision-making in both the arrangement stage and the stages that follow in the composition process.

8. Conclusions and Implications

The key finding from the 'descriptive' study is that attempting to recreate my synaesthetic experiences in the form of computer-generated animations results in an uncanny valley-like effect. I can conclude that these experiments were flawed in the following ways:

1. It appears to be impossible to accurately represent my synaesthetic experiences in the form of computer-generated animation, due to the 'Martian colour' effect (Ramachandran and Hubbard, 2001).

2. Any animations produced by me that describe my synaesthetic visualisations of accompanying music may clash with viewers' own visual interpretations of the music.

It became apparent that, if I were to use my synaesthesia to translate between sonic and visual mediums as inspiration for new creative works, these works should be musical in nature. The key findings from the 'analytic' study are as follows:

1. A process of audiation must be used in order to reverse-engineer my auditory-visual synaesthetic experiences, when translating a visual image into a musical work.
2. Discernible forms in the visual image must be interpreted as being in motion, in order for a visual stimulus to be translated into a musical work. (Otherwise, any audiations will only be of sounds with fixed timbre and pitch, and lacking in rhythm and musical progression/development.)
3. Accurately and precisely translating (i.e., translating all aspects of) a visual image into music results in an unimaginative composition process involving minimal creativity and artistic input.

I hypothesised that the third point could be remedied by choosing only certain aspects of a visual image to translate into musical ideas that could then form the basis of a new composition. I suspected that doing so would allow for a more creative approach to music-making. The key findings from the 'aesthetic' study are as follows:

1. Choosing only certain elements of a visual stimulus to translate into a sonic palette for use in a musical work affords me more artistic control and creative input in the composition process than translating an entire pre-existing visual stimulus into a musical composition.
2. A back-and-forth translative approach between visuals and sound can be employed to provide further inspiration in the sound design stage of the composition process.
3. When in the arrangement stage of the composition process (i.e., when working on the structure and texture of the composition), the chosen combination and sequence of musical ideas is at least partially determined by the extent of my appreciation for the resulting synaesthetic experience.

Overall, while my 'descriptive' and 'analytic' studies did not initially prove to be fruitful, the 'aesthetic' study that followed led to the development of a crossmodal translation-based approach to music composition that I intend to carry forward into future work, and the creation of several compositions of which I am particularly proud.

This paper highlights how investigating the impact of one's synaesthesia on their music composition processes can bring substantial benefits both to these processes and to the resulting music outputs. As I have demonstrated, synaesthetic experiences can provide the synaesthetic composer (or artist, more generally) with an endless source of inspiration and enable them to develop their own unique artistic identity. Although this research focuses on my own experiences and practice and, as a result, findings cannot be universally applied to all synaesthetic artists, I hope that my research will encourage others to share their experiences, in order to learn more about how synaesthesia affects their own creative practice, to develop new artistic methods involving their synaesthesia, and to raise wider awareness of the phenomenon. Due to the idiosyncratic nature of the phenomenon, further research into other synaesthetic artists' practice would provide a fuller and more accurate picture of the role synaesthesia can play in creative practice. Future studies could, for example, investigate the prevalence of the previously unexplored phenomenon of synaesthetic reverse-engineering in synaesthetic artists' creative practice.

Since the neurological and psychological explanations of synaesthesia are beyond the scope of this research, I have not investigated the causes of my synaesthetic experiences or proposed an explanation for my auditory-visual synaesthetic correspondences. Auditory-visual synaesthetic correspondences, particularly those not previously studied in depth (e.g., timbre-texture and timbre-state of matter), could be explored in future neurological and psychological research.

This paper contains material adapted from my PhD thesis *A Compositional Exploration of Auditory-Visual Synaesthesia* (Anderson, 2023).

References

Abbado, A. (2017). *Visual Music Masters: Abstract Explorations: History and Contemporary Research*, Milan, Skira Editore.

Anderson, C. (2023). *A Compositional Exploration of Auditory-Visual Synaesthesia* (PhD thesis), Edinburgh, Edinburgh Napier University, available online at https://doi.org/10.17869/enu.2023.3170304 [accessed August 2023].

Asher, J.E. and Carmichael, D.A. (2013). The genetics and inheritance of synesthesia, *The Oxford Handbook of Synesthesia*, eds Simner, J. and Hubbard, E.M., Oxford, Oxford University Press, pp. 23–45.

Bernard, J.W. (1986). Messiaen's Synaesthesia: The Correspondence between Color and Sound Structure in His Music, *Music Perception*, Vol. 4, No. 1, pp. 41–68.

Cavallaro, D. (2013). *Synesthesia and the Arts*, Jefferson, McFarland & Company, Inc..

Chang, H. (2008). *Autoethnography as Method: Developing Qualitative Inquiry*, Oxon, Routledge.

Collins, H. (2010). *Creative Research: The Theory and Practice of Research for the Creative Industries*, Lausanne, AVA Publishing SA.

Colman, A.M. (2008). *A Dictionary of Psychology*, 3rd edition, Oxford, Oxford University Press.

Cytowic, R.E. (2018). *Synesthesia*, Cambridge, MA, MIT Press.

Day, S.A. (2013). Synesthesia: A first person perspective, *The Oxford Handbook of Synesthesia*, eds Simner, J. and Hubbard, E.M., Oxford, Oxford University Press, pp. 903–923.

Gordon, E.E. (1999). All about Audiation and Music Aptitudes, *Music Educators Journal*, Vol. 86, No. 2, pp. 41–44.

Harrison, J. and Baron-Cohen, S. (1994). Synaesthesia: An Account of Coloured Hearing, *Leonardo*, Vol. 27, No. 4, pp. 343–346.

Hubbard, T.L. (2010). Auditory Imagery: Empirical Findings, *Psychological Bulletin*, Vol. 136, No. 2, pp. 302–329.

Lee, C.H. (2018). *Synaesthesia Materialisation: Approaches to Applying Synaesthesia as a Provocation for Generating Creative Ideas within the Context of Design* (PhD thesis), London, Royal College of Art, available online at https://core.ac.uk/download/pdf/196592014.pdf [accessed October 2020].

Moholy-Nagy, L. (1925). *Composition A XXI* (oil on canvas), available online at www.wikiart.org/en/laszlo-moholy-nagy/composition-a-xxi-1925 [accessed January 2024].

Mori, M. (2012). The Uncanny Valley, *IEEE Robotics & Automation Magazine*, Vol. 19, No. 2, pp. 98–100.

Nelson, R. (2013). *Practice as Research in the Arts: Principles, Protocols, Pedagogies, Resistances*, Basingstoke, Palgrave Macmillan.

Nowness (2013). A Synaesthetic Sunday Roast, *Nowness* (website), available online at www.nowness.com/topic/dominic-davies/a-synaesthetic-sunday-roast [accessed July 2021].

Püschel, A. (2017). *Layers of Reality: Perception Study of a Synaesthete*, Breda, The Eriskay Connection.

Ramachandran, V.S. and Hubbard, E.M. (2001). Synaesthesia – A Window Into Perception, Thought and Language, *Journal of Consciousness Studies*, Vol. 8, No. 12, pp. 3–34.

Rudenko, S. and de Córdoba Serrano, M. J. (2017). Musical-Space Synaesthesia: Visualisation of Musical Texture, *Multisensory Research*, Vol. 30, No. 3, pp. 279–285.

Seaberg, M. (2011). *Tasting the Universe*, Pompton Plains, New Page Books.

Shiraga, K. (1970). *Sorin* (oil on canvas), available online at www.artnet.com/artists/kazuo-shiraga/sorin-qjO62FWBX8ZDcgaeYRcRKw2 [accessed May 2020].

Simner, J., Mulvenna, C., Sagiv, N., Tsakanikos, E., Witherby, S.A., Fraser, C., Scott, K., and Ward, J. (2006). Synaesthesia: The Prevalence of Atypical Cross-Modal Experiences, *Perception*, Vol. 35, No. 8, pp. 1024–1033.

Steen, C. and Berman, G. (2013). Synesthesia and the artistic process, *The Oxford Handbook of Synesthesia*, eds Simner, J. and Hubbard, E.M., Oxford, Oxford University Press, pp. 671–691.

Teichmann, A.L., Nieuwenstein, M.R., Rich, A.N., Gauthier, I., and Enns, J.T. (2017). Digit-Color Synaesthesia Only Enhances Memory for Colors in a Specific Context: A New Method of Duration Thresholds to Measure Serial Recall, *Journal of Experimental Psychology: Human Perception and Performance*, Vol. 43, No. 8, pp. 1494–1503.

van Campen, C. (1999). Artistic and Psychological Experiments with Synesthesia, *Leonardo*, Vol. 32, No. 1, pp. 9–14.

van Campen, C. (2010). *The Hidden Sense: Synesthesia in Art and Science*, London, The MIT Press.

Vanhalen, M. (2020). When audio is always visual, *Music and Synesthesia: Abstracts from a Conference in Vienna, scheduled for July 3-5, 2020*, eds. Jewanski, J., Day, S.A., Siddiq, S., Haverkamp, M., and Reuter, C., Münster, WWU Münster, Münster, pp. 67–68, available online at https://re.public.polimi.it/retrieve/handle/11311/1142795/530166/Riccò_DeCordoba_2020_Music%20and%20Synesthesia.pdf [accessed July 2020].

Wannerton, J. (2015). James Wannerton, *Teknèsyn: Synesthesia and Visual Arts*, Fundación Internacional Artecittà, pp. 30–31, available online at https://books.google.co.uk/books?id=9b8GCQAAQBAJ [accessed October 2020].

Ward, J., Moore, S., Thompson-Lake, D., Salih, S. and Beck, B. (2008). The Aesthetic Appeal Of Auditory-Visual Synaesthetic Perceptions in People without Synaesthesia, *Perception*, Vol. 37, No. 8, pp. 1285–1296.

Weiner, J. (2017). The Return of Lorde, *The New York Times Magazine* (website), available online at www.nytimes.com/2017/04/12/magazine/the-return-of-lorde.html [accessed August 2021].

Discography

Bicep (2021). *Isles*. Available at: Spotify (Accessed: 13 April 2023).

Fake, Nathan. (2020). *Blizzards*. Available at: Spotify (Accessed 13 April 2023).

Hopkins, Jon. (2018). *Singularity*. Available at: Spotify (Accessed: 13 April 2023).

Rival Consoles (2018). *Persona*. Available at: Spotify (Accessed: 13 April 2023).

17

REFRAMING CONFLICTS BETWEEN SYSTEMATIC PRODUCTION, CREATIVE PRODUCTION, AUTHORSHIP AND OWNERSHIP

Robert Wilsmore

1. Introduction

This chapter explores recent research into the systematic generation of billions of tunes (sequences of notes) and their relationship to contemporary understanding of authorship, ownership and originality. Drawing on research experiments with co-researchers Robert Wilsmore, Christopher Johnson and mathematician Philip Brady into the systematic note sequence generation of millions of melodies (Wilsmore and Johnson, 2022), and that of Damien Riehl and Noah Rubin in their *All The Music* project that brute forced billions of note sequences (Riehl, 2019), this paper identifies gaps between ontological considerations of music, in particular through the current return to the object through Object-Oriented Ontology (OOO) as set out by Graham Harman (2018), the legal use of existing work as set out by the US Copyright Office (2021), and that of the understanding of artists such as is exemplified by Ed Sheeran's notion of the 'available building blocks' of music (Geraghty, 2023). Within these three strata conflicts arise between physical existence (such as billions of tunes stored on a hard drive) and claims to authorship, for example where an artist claims authorship of a tune that may pre-exist through the act of systematic brute forcing prior to the artistic creation of it. By way of philosophically engaging with these conflicts, a hypothetical construct, 'the great copyright fight', moves the argument to the logical conclusion that if all possible music that could exist does exist, then the notion of copyright itself, certainly that of future claims of copyright, requires a radical overhaul.

2. The Great Copyright Fight

The Great Copyright Fight is a hypothetical conflict where the lawyers of the estates of all the long-dead rock stars battle all the living songwriters for the right to create and own music (put simply, it reduces the current protracted and ongoing process down to a single conceptual construct). Will the legal occupiers of the musical territory hold on to their claim for ownership of that land, or will they have to accept that it is common ground and share it with everyone? Will they have to accept that it was never their land in the first place, however many flags and signifiers they placed upon the location? It is unclear how this hypothetical battle will end but, *reductio ad absurdum*, if we imagine an eventual state of deproduction (Wilsmore and Johnson, 2022) where all possible music, and not just all melodies (Riehl and Rubin have already done that) has been produced and hence all

DOI: 10.4324/9781003396550-17

music exists so that generating any 'new' music is impossible, what remains is either the possibility that no one writes music anymore (and hence have nothing to own), or composers accept that their so-called 'new' song is in existence already (and hence cannot own it). That nothing is new is the postmodern mantra of recent decades past and, much further back, of Parmenides' 'On Nature' from a pre-Socratic era two and a half thousand years ago (pre-existence is not a new concept), but the acceleration of generation through both human and Generative Artificial Intelligence (GAI) activity means that there might be a physical reality to the existence of all music rather than only a philosophical one, which highlights a current gap between the law and ontology, and lifts the outcome of the great copyright fight out of the world of hypothesis and into a that of a distinctly possible reality.

3. All the Music (ATM)

As part of their *All The Music* project (Riehl and Rubin, 2023) Damien Riehl and Noah Rubin have copyrighted billions of tunes in America in accordance with definitions of what constitutes legal existence of a tune for copyright purposes, by storing them all in midi format on a hard-drive meeting the criteria of being "fixed in a tangible form" (US Copyright Office, 2021). By legal definition, they exist although ontologically it is less clear that the digital zeros and ones of the storage on the hard drive constitute existence any more than an algorithm that can recreate the tune on demand. But the latter would unlikely be accepted as legally existing because it is not in a "fixed and tangible" format, even though (as Riehl has pointed out) the reproduction from the algorithm is a mathematical certainty. A comparative example here is that where ATM tunes are in 'fixed' files, Jonathan Basile's *The Library of Babel* project (Basile, 2015), which 'contains' all possible texts, and hence contains all possible song lyrics that could be written, uses an algorithm to reproduce the text from its mathematical location on demand because there is simply not the storage capacity to store the texts in the way that ATM does. Hence, in the US at least, the files from ATM are likely copyrightable but the song lyrics in *The Library of Babel* are not. The law attempts definitions of existence for pragmatic purposes but it is not the same thing as existence outside of the law itself, that is, the *thing-in-itself* and our relationship with it. The dilemma between creativity and copyright are demonstrated time and again in news and social media. For example, songwriter Arie Burshtein convincingly argues that "copyright is one of the building blocks of freedom and – essentially – human rights" (Burshtein, 2023). But in an other very public case, Ed Sheeran said he would quit music if he lost the copyright trial relating to 'Thinking out Loud' (Geraghty, 2023). The law, to his mind, would have taken away his freedom to compose. Copyright seems to be both an essential component of freedom *and* a toxic inhibitor of it.

So, what are the possible responses to this? There is the objective approach of forensic musicologists like that of Professor Joe Bennett at Berklee College of Music (see Bennett, 2023) who has developed a systematic approach to comparing similarities that seeks to demonstrate levels of difference and similarity between songs. Bennett, quoted in the *Guardian* newspaper (Khomami, 2022) on the Ed Sheeran 'Shape of You' case, said that "a common musical error that listeners can make [. . .] is to assume that plagiarism is the only explanation for one melody being slightly similar to another", with the article's writer Nadia Khomami commenting that in today's age of streaming "how difficult it is to differentiate between coincidence, inspiration and theft" (Khomami, 2022). And on the 'Thinking out Loud' case, in an interview with Victor Blackwell for CNN, Bennett showed his support for Sheeran's creativity, saying that "he's standing up for songwriters everywhere by sticking to his integrity and [. . .] holding out for the truth" (CNN, 2023). If music has the right to children then forensic musicology is the DNA test to identify who the parents are. But does music have the right to children? *All The Music* thinks it does, taking an alternative approach, systematically brute forcing all possible tunes, copyrighting them, and placing them in the public domain through the Creative Commons organisation who "advocate for *better sharing*: sharing that is contextual, inclusive, just, equitable, reciprocal, and sustainable" (Creative Commons, 2023). Creative Commons discussions seem well-balanced; on whether GAI has vicarious liability with regard to copyright, Stephen Wolfson discussed the question "should GAI tools be held responsible for potential copyright infringement conducted by users of these tools?" and concluded that "we need to figure out norms and best practices that can allow these promising new technologies to develop and thrive, while also respecting the rights and concerns of artists and the public interest in access to knowledge and culture" (Wolfson, 2023).

In practical terms, *all* tunes are now copyrighted and are, arguably, available to use, and Riehl and Rubin have made a direct challenge to the legal system in their attempt to allow creativity to bloom by "helping songwriters make all of their music" (Riehl and Rubin, 2023). It does not prevent cases of plagiarism coming to court but it does strengthen one part of the argument by "copyrighting all the melodies to avoid accidental infringement" (Riehl, 2019). In fact, with all tunes now 'existing', composers can no longer accidentally stumble on an existing tune because, with almost absolute certainty, they *will* be using an existing tune. Ed Sheeran said, "It is my belief that most pop songs are built on building blocks that have been freely available for 100s of years" (Geraghty, 2023). Should we now consider all tunes as being freely available building blocks, because they are all now in the public domain? It is yet to be seen how this plays out in the courts though, and given it is not similarity itself but *previous knowledge* that is often the issue, such as in 'Thinking out Loud' and the 'Shape of You' cases, protection from accidental infringement will only be one part of the argument, albeit an important one.

4. Toast Theory

Another way to address this fundamental issue of freeing, rather than curbing, creativity is to challenge and change societal perceptions and widely held ideologies of plagiarism. In an article in the *Leonardo Music Journal* in 2002 on 'Pleasure' I posed as a research question "Is there really a difference between the way we enjoy classical and popular music" (Wilsmore, 2002) and addressed this by composing and directing the same piece (*Technomass*) sung by a choir that was *a cappella* in one version and accompanied by dance based (EDM) backing in another. Later, in our mathematical research that became abbreviated to 'Toast Theory', the question was changed to "is there really a difference between the way we *listen* to classical and popular music?" and we found that the answer is "yes, there is" and it was not a positive finding either. 'Toast' is a corrupt acronym of 'The Song of a Thousand Songs' (Wilsmore, 2014) which is the name given to the (hypothetical) *only* song that exists, all 'songs' are just continuous parts of this one song. So-called individual songs do not have the endings we think they have, we are all collaborating on just the one great big song. Toast is a neo-hippie monism, but it is not a spaced-out, drug-induced, spiritual connection to the universe, it has a function, and that is to clarify, and then change, how we perceive similarities in music. Where ATM is a challenge to the legal system as well as an encouragement to create, Toast is a challenge to an ideology. In brief, Toast states:

> Musical phrases in which we perceive resemblance operate through the same modes of relationship regardless of whether they occur within one piece of music or in separate pieces of music.

(Wilsmore and Johnson, 2022, p.159)

FIGURE 17.1   Two similar phrases in a 'singular' piece of music.

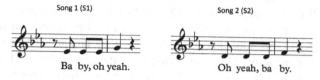

FIGURE 17.2   Two similar phrases in 'separate' pieces of music.

And continues:

> When we hear a theme develop in a symphony, we heap praise on an author. When we hear two similar themes in different songs, we heap criticism and shouts of 'plagiarism' upon one of the authors. Yet, in terms of what we hear, there may be no difference at all in these two scenarios.
>
> *(Wilsmore and Johnson, 2022, p.159).*

Hence, when Beethoven follows his theme A1 with the theme A2 (figure.17.1) we are inclined to give positive praise to him as a genius for his creative abilities to develop A1 into A2. But when we hear one tune (S1) and then soon after we hear another song with a similar tune (S2) (figure.17.2) we are inclined to give *negative* criticism of the second author. Not the positivity of creative development, but the negativity of plagiarism and lack of creativity. Yet there may be very little, or no difference, in the two listening experiences. These positive or negative perceptions are about the author, not the sound. Sonically (melodically at least), the examples are more or less the same, the differences are non-sonic signifiers i.e., nothing to do with the sound (e.g. the name of the composer, the title of the song, the end of the playlist, etc.). Signifiers are not eternally fixed, they may be changed, they may be "deterritorialized" or "annulled" (Deleuze and Guattari, 1987). The law puts up its signs, its signifiers of ownership, but in society these are not always accepted as fact. In the hypothetical great copyright fight we might view the two sides as standing on either side of Woody Guthrie's signpost, where one side states ownership of the territory whilst the blank reverse side is taken as meaning that the territory is common ground. In reality, we spread this binary (the opposing sides of the sign) into a spectrum, and in doing so create the problem of where to place the boundaries, hence the never-ending arguments about where to draw the lines. Put simply, the law wants boundaries but spectra don't have them. One thing we can do is to nullify the boundaries and perceive music as one song, one symphony, one huge set of variations. Boundaries are a mindset, an ideology, they can be moved or dismantled. Hence Toast's manifesto is:

> As a society we should perceive similarities as signs of positive influences on creativity, not as negative accusations of a lack of creativity.

An obvious line of flight in the exploration following on from the Beethoven's 5th thematic development was to systematically write every possible tune that could exist as a step towards the state of deproduction, of the end of music production where there is no more to produce. In 2017, mathematician Phil Brady and myself sought to produce every possible tune within the basic parameters of the opening two-bar phrase of Beethoven's 5th Symphony. If Beethoven was writing a set of developments on his 'da da da dah' theme, then what was the complete mathematical 'tune set' that he could have drawn from? We included the pitch-set within an octave, rhythm and rests between 8th and whole notes (including the rests was trickier than expected and required some complex mathematical partitioning). The resulting matrix, even with such limited parameters, produced over 83 million tunes (see Wilsmore and Johnson, 2022, pp.190–199). This exercise was part of a much larger work on Coproduction, a book which was delayed by Covid and was not published until 2022. By which time Riehl and Rubin had released their TEDx talk in 2019 having brute forced billions of pitch combinations at the time. On seeing what they had achieved I did briefly have a moment of dread, thinking "Help! I'm going to be sued on 83 million accounts of plagiarism!" But Riehl's intention is to *allow* creative freedom not to *limit* it. In ATM they argue that melody should be uncopyrightable, the argument is logical and convincing, they mathematically brute forced billions of pitch combinations, hence melody is maths, and because maths is facts and facts cannot be copyrighted, melody cannot be copyrightable.

MELODY = MATH = FACTS = UNCOPYRIGHTABLE

(Riehl and Rubin, 2023)

Neither ATM or Toast theory want to play into the hands of real thieves either, so it is not a call to abandon copyright, just to remove the toxic inhibitors and release the freedom.

5. Asymmetrical Collaboration

Toast theory breaks down the semiotics of boundaries between individual songs and by doing so suggests that there is only one song and that we are all collaborating on, namely 'The Song of a Thousand Songs', and this aligns with Robert C Hobbs' proposition in his essay 'Rewriting History: Artistic Collaboration since 1960' that "collaboration is, in essence, nothing more or less than influence positively perceived as part of an on-going cultural dialogue" (Hobbs in McCabe, 1984, p.79), and we can simplify this to "Collaboration is influence". But collaboration would normally be considered a two-way exchange. For example, B receives an idea from A, B develops the idea and gives it 'back to' A, and the process can repeat.

A ---> B
B ---> A
[repeat]

An end product of this process, e.g. a song, can be considered to be a collaboration. But what if the process does *not* have this reciprocal A-B symmetry? Let's assume A is a long-dead rock star, and B a living songwriter. B receives an idea from A and develops the idea. They cannot return it to A because A is dead:

A ---> B
B ---->X A

Normally we would then say that B has not collaborated with A but has been 'influenced' by A (or worse still that they 'plagiarised' A) and with regard to influence Harman puts it that "there is no such thing as reciprocity; influence is never mutual, but always leads in just one direction" (Harman, 2010, p. 96). Similarly, collaborations are not normally equal and symmetrical. Collaborations are often done in 'series', for example, here are three typical collaborative modes of operation:

> **Parallel** – where contributors work separately on a project but within the same timeframe
> **Joint** – where contributors are 'present' together for immediate responses.
> **Series** – where a contribution is 'handed on' to the next contributor.
>
> *(Alix et al., 2010)*

In Series type collaborations A makes their contribution, then B takes this on and makes their contribution which is then given to C, *etc.* The key point is that B does *not* necessarily need to return their contribution to A in order for it to be a collaboration. In copyright law B might require A's *permission* to develop their idea but I am referring to collaborative processes here, not to copyright, and there is no *explicit* rule in collaboration that requires B to have A's permission. It poses a fundamental question for us to answer: Is permission a predicate of collaboration? If it is, then it might seem that we can draw a clear distinction between collaboration and influence, but as influence may be either consensual or non-consensual, and as collaboration includes influence within its processes, then it may not get us any further towards a useful understanding of the word.

There are many terms for the Series type of 'handed-on' or 'picked-up' construction of a work that does not have some sort of explicit agreement between contributors. In Toast theory it is referred to as "Collaboration without consent – Denial or unknowing collaboration", Keith Sawyer calls it "Invisible collaboration", Brian Eno might refer to it as "Scenius", Robert C Hobbs as "Influence" and "ongoing cultural dialogue", to Deleuze and Guattari it is a "Rhizome", to Bennis and Biederman it is part of "Organised Genius" (see Wilsmore and Johnson, 2022 pp. 86–87). Triple O, or Object-Oriented Ontology, philosopher Graham Harman uses the term "asymmetrical or vicarious causation" (Harman and Wier, 2021) where the interaction of A-B is not equal or symmetrical, as opposed to Newtonian causality where there is symmetry, i.e., "every action has an equal and opposite reaction". In OOO, asymmetry does not tilt the balance nearer to an unknowable truth but instead adds layers, "not digging downward but *building upward* to a higher layer" (Harman, 2018, p.83). In a conversation with Simon Wier on OOO and architecture, Harman says on asymmetrical causation that "I as a human can observe long-dead stars, which thereby have an effect on me even though I can't possibly affect them in turn" (Harman and Wier, 2021). Harman's literal long-dead stars can affect but not be affected in return. And the same goes for the long-dead rock star who can affect but not be affected. Weir adds that "asymmetrical causation is perception" (Harman and Wier, 2021) and this perception can be generative, in that is that it goes on to create things. We might term it as influence of the long-dead star or the long-dead rock star, but there is no influence until something is created that contains the influence (causation). Harman adds that "there is always a point at which sensual objects can become real, often through the mechanism of social acceptance" (Harman and Wier, 2021) echoing Hobbs' notion of cultural dialogue. The Song of a Thousand Songs, through the mechanism of social acceptance, becomes a real object, and if the variations within this song (i.e., all the so-called individual songs) are perceived as the positive generations of asymmetrical collaboration then as a society we will perceive similarities as signs of positive influences on creativity, not as negative accusations of a lack of creativity.

6. Conclusion

The Great Copyright Fight may only be a conceptual vehicle with which to reduce a real process that will last many decades or even centuries into a single manageable idea, but in forcing us to consider what we will do when all music already exists it can affect our reality now. So why not consider that we are already in a state of absolute deproduction, of everything already existing now, just as Parmenides did two thousand five hundred years ago? It will not be an anarchic *free-for-all*, we will still hold on to concepts of creativity and justice. But we will have to, sooner or later, find new ways to consider what creativity and justice are, and in doing so maybe we will find a fairer way with which to deal with authorship and ownership.

Acknowledgments

I am indebted to the conversation, correspondence, and generosity of Damien Riehl in helping to articulate some of the arguments presented here.

References

Alix, C., Dobson, E., and Wilsmore, R. (2010). *Collaborative Arts Practices in HE: Mapping and Developing Pedagogical Models*. Palatine, The Higher Education Academy.

Basile, J. (2015). Website. *The Library of Babel*. Available at https://libraryofbabel.info/ [Accessed 5 June 2023].

Burshtein, A. (2023). Social Media post 'Music Copyright and Integrity in the Music Industry'. LinkedIn.

CNN (2023). *Joe Bennett in interview with Victor Blackwell*, 30 April 2023. CNN News.

Creative Commons (2023). 'Sharing that is contextual, inclusive, just, equitable, reciprocal, and sustainable'. Website. Available at https://creativecommons.org [Accessed 5 June 2023].

Bennett, J. (2023). Website. *Joe Bennett Music Services*. Available at https://joebennett.net [Accessed 14 January 2024].

Deleuze, G. and Guattari, F. (1987). *A Thousand Plateaus: Capitalism and Schizophrenia*. Trans. Brian Massumi. Minneapolis, MN, University of Minnesota Press.

Geraghty, H. (2023). NME online, 2 May 2023, 'Ed Sheeran warns he'll quit music if found guilty in copyright trial'. Available at www.nme.com/news/music/ed-sheeran-warns-hell-quit-music-if-found-guilty-in-copyright-trial-3438272 [Accessed 5 June 2023].

Harman, G. (2010). 'Asymmetrical causation: influence without recompense', *Parallax*, 16:1, 96–109.

Harman, G. (2018). *Object-Oriented Ontology*. UK and USA, Pelican Books.

Harman, G. and Wier, S. (2021). 'Architecture and Object-Oriented Ontology', Simon Weir in conversation with Graham Harman, isparchitecture.com. AP vol 5, No 2.

Khomami, N. (2022). 'Coincidence or theft? Rise of music streaming can make it hard to judge', *The Guardian*, 6 April 2022.

McCabe, C. (1984). *Artistic Collaboration in the Twentieth Century*. Washington, D.C., Smithsonian Institution Press.

Riehl, D. (2019). TEDx Minneapolis 'Copyrighting all the melodies to avoid accidental infringement'. Available at www.youtube.com/watch?v=sJtm0MoOgiU. [Accessed 5 June 2023].

Riehl, D. and Rubin, N. (2023). All-The-Music.info website. Available at http://allthemusic.info/ [Accessed 5 June 2023].

US Copyright Office (2021). *Sampling, Interpolations, Beat Stores and More: An Introduction for Musicians Using Pre-existing Music*. Washington, Copyright.Gov. Available at www.copyright.gov/music-modernization/educational-materials/Sampling-Interpolations-Beat-Stores-and-More-An-Introduction-for-Musicians-Using-Preexisting.pdf [Accessed 5 June 2023].

Wilsmore, R. (2002). 'Techno, Trance and the Modern Chamber Choir: Intellectual Game or Music to Groove to?', *Leonardo Music Journal*, 12, 61–63.

Wilsmore, R. (2014). 'The Song of a Thousand Songs: Popular Music as Distributed Collaboration.' March 2014 at IFAAI (International Festival for Artistic Innovation) Leeds College of Music, 10th–14th March 2014.

Wilsmore, R. and Johnson, C. (2022). *Coproduction: Collaboration in Music Production.* New York and London, Routledge.

Wolfson, S. (2023). March 24, 2023 'Style, Copyright, and Generative AI Part 2: Vicarious Liability'. Available at: https://creativecommons.org/2023/03/24/style-copyright-and-generative-ai-part-2-vicarious-liability/ [Accessed 5 June 2023].

Discography

Sheeran, E., Williams, J., and Wadge, A. (2014), 'Thinking out Loud'. Asylum, Atlantic Records.

Sheeran, E., McDaid, J. *et al* (2017), 'Shape of You'. Asylum, Atlantic Records.

18

AN INNOVATIVE MUSIC PRODUCTION MODEL LEADING TO A SUSTAINABLE HIT SONG

"Främling"

David Thyrén and Jan-Olof Gullö

1. Introduction

The following chapter presents selected findings from a study that explores the innovative factors behind the success of the Swedish song "Främling" [Love Isn't Love]. The study delves into the background, creation, launch, promotion, aftermath, and significance that "Främling" had for the Swedish music industry. The song was performed by Carola Häggkvist, who was 16 years old at the time and represented Sweden with "Främling" in the Eurovision Song Contest of 1983 (Rylander and Karlsson, 2016).

The purpose of this study is to identify the main factors that led to the success of the song "Främling" and assess their level of innovativeness. The overarching research question is: What decisive success criteria can be identified in "Främling" regarding key personnel, infrastructure, entrepreneurship, marketing, artistic and technical qualities? By conducting an analysis of the song's music, lyrics, and production, as well as reviewing relevant literature and conducting interviews, the key criteria that contributed to the song's popularity have been determined.

This study is part of an ongoing project, "Searching for Sophia in Music Production". The project aims to identify critical factors in successful music production, by exploring various aspects of the music industry and Swedish music exports (cf. Gullö and Thyrén, 2019). Specifically, the study focuses on music production with theoretical perspectives from music education, musicology, and sociology. This research project aims to enhance comprehension of innovative music production techniques and facilitate the development of music production programs in higher education. The term Sophia, meaning wisdom, originates from ancient Greek knowledge typology (Aristotle, 2011). It serves as a summary of the theoretical framework of the project.

As a part of the Searching for Sophia in Music Production project, a previous study has identified several influential individuals in the Swedish music industry who possess unique innovative abilities (Gullö et al., 2022). These individuals have shown a tendency to break away from traditional norms and instead follow their instincts, leading them to explore new and uncharted paths. They are referred to as "Icebreakers". As their work leads them to explore new paths, their actions create opportunities for others to follow and build successful teams and companies. These successful teams and companies are referred to as "Clusters". Such Icebreakers and Clusters have significantly contributed to the growth of Swedish music exports. Over the years, they have surpassed expectations and opened new opportunities for the Swedish music industry. The

DOI: 10.4324/9781003396550-18

following Icebreakers and Clusters have been identified: *Kurt Atterberg* (1887–1974) was a well-known composer and civil engineer who founded STIM – a non-profit organisation that ensures fair compensation for musicians and publishers (Gullö et al., 2022); *Stikkan Anderson* (1931–1997), is renowned for his work as ABBA's manager and primary lyricist and for establishing the Polar Music Prize (Gradvall, 2023; Hedlund, 1983); *Bert Karlsson* (b. 1945) is a successful entrepreneur who established Mariann Records and, while not reaching the same level of global triumph as his inspiration Stikkan Anderson, he once held roughly a third of the record market in Sweden (Rylander and Karlsson, 2016); *Ola Håkansson* (b. 1945) is a versatile person who gained popularity in the 1960s as the frontman of the popular group Ola & the Janglers. He founded Stockholm Records and Ten Music Group, through which he has promoted many international artists, including A-Teens, Zara Larsson, and Icona Pop (Gullö and Thyrén, 2018); *Robert von Bahr* (b. 1943) is a successful music producer and executive at BIS Records, specialising in classical music with a focus on Scandinavian genres. He founded BIS in 1973 and has promoted exceptional Swedish classical musicians for 50 years, earning him numerous awards and making him a key part of Swedish music exports (Arvidsson, 2007; Östman, 2018); *Pelle Karlsson* (b. 1950) is a talented musician who through Prim Records in the 1970s, and his Christian progressive style, became popular in Sweden, neighbouring countries, and the US (Gullö et al., 2022); *Denniz PoP*, also known as Dag Volle (1963–1998). He was a successful music producer who mentored Max Martin and worked with artists like Westlife and Britney Spears. He began as a renowned DJ in Stockholm and co-founded Cheiron Productions (Thyrén et al., 2020); *Daniel Ek* (b. 1983) founded Spotify, a popular music streaming service. For many years, *Billboard* (2023) has recognised Ek as one of the top players in the music industry.

As part of the study, Bert Karlsson was interviewed to examine his innovative accomplishments and contributions through Mariann Records, which included the production of the hit track "Främling" (Gullö, 2023).

2. How "Främling" was Created

The following section provides research findings and analyses related to the song "Främling". This includes information about the individuals involved in creating the song, its performance on national and international television during the Eurovision Song Contest, the song's lyrics and music, and its impact and success.

The song "Främling" was created by a team of talented individuals. The team consisted of *Bert Karlsson*, a record label executive; *Lasse Holm*, a songwriter; *Monica Forsberg*, a lyricist; *Lelle Sjöholm*, an arranger; and *Carola Häggkvist*, a performer. It is important to note that many more people were involved in the project, but these individuals were the key figures in its production.

2.1. Bert Karlsson

Bert Karlsson is, in Sweden, a well-known entrepreneur and businessperson who has succeeded in various fields, including Swedish popular music. He was born in 1945 in a rural part of Sweden, Vads församling in Västergötland, and has mostly stayed away from the music industry hub in Stockholm. From a young age, he displayed leadership, initiative, business judgment, persistence, competitiveness, a love for sports and music, and a willingness to take calculated risks. Bert Karlsson values hard work and avoids distractions and health hazards like alcohol and tobacco. He has a unique talent for identifying attractions that may have mass appeal and that people are willing to pay for. Although he primarily works in popular culture and is not spiritual, he has collaborated successfully with many religious artists and musicians.

At age twenty, Bert Karlsson made a significant career advancement by becoming a grocery store manager, marking his first major accomplishment. At this time, Bert Karlsson also began working as a booker for local dance bands in addition to his regular work. The reason behind this was that the girlfriend of Bert Karlsson's wife's brother happened to be the female vocalist of a band known as Magnus Kvintett. Through this family connection, Bert Karlsson became increasingly fascinated by the music scene and tried to help organise and book concerts and tours for local bands. For this purpose, he started a booking agency. Bert Karlsson soon experienced difficulties booking gigs and realised the importance of providing records. That would speed up the bookings and provide better conditions for the bands.

Bert Karlsson rented studios in the Skara area and released recordings on minor independent labels. However, the quality was insufficient, so he went to Stockholm to work with the successful bandleader and music producer Kjell Öhman to produce higher-quality productions. The idea was basically to record Swedish versions of American or English hits.

During this period, in the late 1960s, discotheques were a new phenomenon in Sweden. Bert Karlsson had noticed how discotheques had become popular in Stockholm, and he decided to start his own discotheque in Skara. He ventured out on a small basis in a local restaurant. After selling the grocery store, Bert Karlsson raised the necessary money to start a bigger discotheque in the city of Lidköping, Bert Karlsson made sure to book many dance bands and pop acts to play in his discotheque, hence combining his roles as a booking agent and discotheque owner. He ran a tight ship and saved money by also taking on the roles of disc jockey, administrator, cleaner of the facilities, networker, and negotiator with the local authorities. To avoid legal problems, the discotheque was run as a private club and, consequently, had to change names regularly. One of the names was Mariann, which Bert Karlsson later reused when setting up his record company. Around the same time, he discovered the lucrative business of providing bingo and soon started a bingo hall for a clientele of elderly ladies. This gave him a solid income, and Bert Karlsson decided to stake a serious investment in the music industry.

Mariann Records was launched by Bert Karlsson during the summer of 1972. His business model was simple: working with skilled performers and musicians and producing good-quality recordings with a distinctly recognisable sound. He worked out a distribution deal with EMI but quickly understood the importance of being independent and as soon as possible managed to take control of the distribution through his own record company. Bert Karlsson also began collaborating with the KMH-studio in Stockholm, run by Lennart Karlsmyr, Leif Malmborg and Lasse Holm. Bert Karlsson realised he needed a good producer and started working with music producer Lars O. Carlsson, who used to play with Lasse Holm in the band Moonlighters.

Mariann Records managed to enter the all-important chart of Svensktoppen on the Swedish public service radio. This first breakthrough came with the song "Hallå du gamle Indian" with the Jigs. The song entered at number seven and peaked at number three. Bert Karlsson realised that the musicians in the dance bands he worked with were not technically sufficient to play on professionally made recordings. Therefore, he hired skilled studio musicians. In fact, he invested in working with a short list of the best of Sweden's studio elite, a Wrecking Crew of the same musicians that also appeared on recordings with ABBA and other Swedish high-class acts. He also started experimenting by trying out songs with various artists to see which artist matched each song the best, seeking the ideal balance between artistic performance, perfection, and charm. The method was intended to increase the potential of the songs to become hits on Svensktoppen. These methods were criticised by some as too commercial and industrialised but Bert Karlsson, realising that his methods worked very well, was indifferent towards the critique. Only later did he learn that he applied similar strategies as American hit factories, e.g. demon producer Phil Spector, Phillysound, and record executive Berry Gordy Jr. of Motown.

Having established Mariann Records on Svensktoppen, Bert Karlsson continued to develop his record company and started to employ several in-house composers and music producers. Lasse Holm, from the KMH-studios, and Torgny Söderberg, a musician and composer, were the most prominent. Both achieved great success. Bert Karlsson also discovered and signed new acts, mainly in the dance band tradition. Those bands started to sell albums in millions. One dance band, Vikingarna, sold 13 million albums. Inspired by Stikkan Anderson and his vast international success with ABBA through the Eurovision Song Contest, Bert Karlsson's next move was to promote his products on television. In 1978, Mariann Records started to launch their music in the Eurovision Song Contest, and this long-term investment continued until 2004 (Arvidsson, 2007; Gullö, 2023; Karlsson, 2006; Rylander and Karlsson, 2016; Östman, 2018).

2.2. Lasse Holm

Born in Stockholm, 1943, Lars-Eric "Lasse" Holm is a talented musician with extensive experience in music production, songwriting, singing, and playing various instruments. As a child, Lasse Holm played the piano and the trumpet, and later discovered his passion for rock 'n' roll, which led him to learn the guitar. In the '60s, Lasse Holm was a member of several bands, including the Moonlighters, who achieved moderate success. The Moonlighters toured internationally in 1967 and signed a recording contract with Sonet, a record label that also promoted Lasse Holm as a solo artist named Larry Moon.

In 1970, Lasse Holm co-founded the KMH-studio in Stockholm and a year later, he began collaborating with Bert Karlsson, who later employed him as an in-house producer and composer at Mariann Records. Lasse Holm also performed occasionally as an artist during his whole career.

Lasse Holm has been a frequent songwriter and performer in the Swedish Eurovision Song Contest, with an impressive 18 contributions, including an incredible five wins (1982, 1983, 1985, 1986, and 1993). "Främling" was his sixth entry in the Swedish Eurovision Song Contest, and the second time he won as a composer (Holm and Lundin, 2015; Löfström and Eklöf, 2019).

2.3. Monica Forsberg

Monica Forsberg, a multi-talented artist, was born in Karlskoga in 1950. She has excelled in various fields such as singing, songwriting, acting and film production. Monica Forsberg has dubbed several Disney movies and cartoons. During the '80s, she was a member of the band Ritz, and participated as a performer in the Swedish Eurovision Song Contest in 1983 and 1985. However, her most significant impact came as a lyricist, frequently collaborating with Lasse Holm as a songwriter. She contributed twelve songs, including two wins in 1982 and 1983. "Dag efter dag" [Day After Day], written in collaboration with Lasse Holm and performed by Kikki Danielsson and Elisabeth Andreasson as Chips, won in 1982. In 1983, Monica Forsberg wrote the lyrics for "Främling" (Löfström and Eklöf, 2019).

2.4. Lelle Sjöholm

Lennart "Lelle" Sjöholm, the arranger of "Främling", was born in Jönköping in 1945. He is a versatile musician and music producer who has collaborated with numerous prominent Swedish artists as an arranger and music producer, including Kikki Danielsson, Carola Häggkvist, Christer Sjögren, Lisa Nilsson, Tommy Körberg, Sanna Nielsen, Jan Malmsjö, and Arja Saijonmaa. Throughout his long and successful career, Lelle Sjöholm has been awarded over 100 Gold-Platina records. He even received a Grammy Award for collaborating with the American gospel singer Andraé Crouch (Löfström and Eklöf, 2019).

2.5. Carola Häggkvist

Carola Häggkvist, also known simply as Carola, was born in Stockholm in 1966. She is a very talented performer who discovered her passion for the arts and pursued it as a career at a young age. Carola received a comprehensive music education, attending Maria's School, Adolf Fredrik's Music School, and Södra Latins Music Gymnasium. As a child, she participated in various talent shows and competitions. In December 1981, Carola was selected by a jury that included Stikkan Anderson to appear on the immensely popular Swedish TV show *Hylands Hörna*, which is similar to the American show *I've Got a Secret*. Bert Karlsson, who saw Carola on TV, quickly signed her to Mariann Records through his connections with Lennart Hyland, the TV show host. Although Carola was young, she was determined to maintain her musical integrity. As a result, she was put on hold for a year or so while being carefully prepared for her big launch at just the right time (Alrup and Andström, 1983; Ekström and Koljonen, 2012).

2.6. Launching "Främling"

During the late 1970s, Sweden had the highest record sales per capita globally, with most consumers aged between 9 and 24 years. At that time, several debates existed about how the music industry influenced young listeners' musical tastes and behaviours (Brolinson and Larsen, 1981). Bert Karlsson was inspired by Stikkan Anderson, who had achieved immense success with ABBA in the Eurovision Song Contest. This motivated Bert Karlsson to promote his music and artists on television. However, during the 1970s, Sweden only had two public service television channels and no commercial broadcasting. To overcome this, Mariann Records began releasing their music in the Swedish Eurovision Song Contest. In 1978, Mariann Records achieved a shared second position with two songs: "Miss Decibel" by Wizex and "Växeln hallå" by Janne Lucas Persson. Mariann's first victory came in 1982 with the song "Dag efter dag" with the duo Chips, but it didn't gain international recognition. The following year, Bert Karlsson and his two principal songwriters and music producers, Lasse Holm and Torgny Söderberg, made extra efforts to win again and consolidate their success. They worked on many new songs, including "Gubben i lådan" [Jack in the Box]. Carola sang on the demo, and Kikki Danielsson, the more experienced performer, sang on the final version, but the song didn't qualify for the Swedish Eurovision Song Contest competition.

In the meantime, Lasse Holm devoted his efforts to composing "Främling". Both he and his colleague Torgny Söderberg were highly productive. They would spend long hours in the studio, constantly writing and recording new songs. Lasse Holm has, over the years, released over 700 songs. Their employer, Bert Karlsson, wanted powerful titles for each song. When Lasse Holm came up with "Främling", he composed it on the piano and recorded a demo version on a portastudio at home. He sang the song himself in English, with unfinished lyrics. The recording quality was mediocre, but the song's essence was present, with the crucial words Stranger and Mona Lisa in the chorus. The demo inspired Monica Forsberg to develop the final lyrics. The next step was to record the song professionally, with some of Sweden's most prominent studio musicians, including Peter Ljung on keyboards and Håkan Mjörnheim on guitar, along with a rhythm section consisting of ABBA musicians Rutger Gunnarson on bass and Per Lindvall on drums. Lasse Holm and his team worked hard to make it a polished and professional music production (Löfström and Eklöf, 2019).

At first, Lasse Holm was considered to perform the song, but Forsberg's lyrics made it more suitable for a female performer. Bert Karlsson searched for several artists and even considered Kikki Danielsson, but she was chosen for another song called "Varför är kärleken röd?" [Why is love red?] by Torgny Söderberg. For the Swedish Eurovision Song Contest, an anonymous studio singer was used by Mariann for "Främling". However, Carola expressed interest in the song and was allowed to audition. Her performance in the studio was outstanding, and Bert Karlsson was impressed by how well the artist matched the song. Despite the risk, Bert Karlsson chose the unknown 16-year-old Carola as the performer for this significant song.

In 1983, the Swedish Eurovision Song Contest saw the simultaneous launch of "Främling", and Carola as a new and upcoming performer. The success was extraordinary and still resonates in Sweden today. Carola won by a landslide, with the highest points from all eleven jury groups. Mariann Records also secured the second position with Kikki Danielsson singing "Varför är kärleken röd?" In the international Eurovision Song Contest final in Munich, "Främling" made a considerable impact, reaching the third position with 126 points, despite being sung in Swedish, a language not understood by most of the international audience. Over 80% of the Swedish population watched the event on TV.

2.7. The Song

"Främling" has a catchy "instrumental" hook that starts with a short descending phrase in the key of F minor. The phrase ends with a trill on a minor second, which gives the song a slightly oriental feel that matches the theme of a mysterious stranger. The song follows the structure of an intro, two verses, a chorus, an intro reprise, a third verse, two choruses, and an outro. The song features key changes that shift between minor and major keys. The chorus is uplifting in the parallel major key of F major and has a positive vibe that contrasts with the descending and mysterious character of the verses. "Främling" ends with increased uplifting energy as the arranger Lelle Sjöholm focused on making the last few seconds truly spectacular with a musical fanfare perfectly synchronised with a final striking winning pose by the performer. The lyrics of the first verse are as follows: "Främling – Vad döljer du för mig? I dina mörka ögon – En svag nyans, av ljus nånstans. Men ändå, en främling – Så känner jag för dig. Jag ber dig, låt mig få veta Vem vill du vara? Kan du förklara det för mig?" (Forsberg and Holm, 1983), translated to English: *Stranger – What are you hiding from me? In your dark eyes – A faint shade, of light somewhere. But still, a stranger – That's how I feel about you. Please let me know. Who do you want to be? Can you explain it to me?*

Bert Karlsson took steps to ensure that "Främling" had the potential to become an international hit. He provided text versions of the song in German, Dutch, and English, respectively called "Fremder", "Je Ogen Hebben Geen Geheimen", and "Love Isn't Love". Carola then provided vocals in these different languages to the original musical backing track. For the English version, the renowned musician and music industry executive Stuart Slater (1946–2023) was engaged. The lyrics of the first verse of Slater's English version are as follows: Stranger, you came into my life, touched me with sweet sensation. And so every night, we made it right forever. And stranger, you cut me like a knife, you showed me my destination. I thought I knew love, but then a true love came my way (Forsberg et al., 1983).

2.8. Aftermath

"Främling" became an instant hit in the Scandinavian countries, sparking a long-lasting Carola fever and laying the foundation for Carola's highly successful career. For 40 years, she has remained one of Sweden's and Norway's most famous artists. Her debut album, also titled *Främling*, sold over one million copies and is one of the ever-best-selling national albums in Sweden.

Bert Karlsson's record company, Mariann, experienced significant growth, and success thanks to this sustainable hit. With numerous contributions to the Swedish Eurovision Song Contest and consistent qualifications throughout the '80s, Mariann Records established itself as a major player in the Swedish music industry. Over the span of 26 contributions, including 13 wins, from 1978 to 2004, Mariann dominated the Swedish Eurovision Song Contest during the mid- to late '80s. Emphasising quantity, Mariann had two to three accepted contributions each year, sometimes more. In 1987, they culminated with six accepted contributions, including the winner. Mariann also won twice in the Norwegian Melodi Grand Prix. Despite some entries not qualifying, many of Mariann's songs became hits regardless.

In both 1983–84 and 1991, Mariann accounted for around 35% of record sales in Sweden. This helped to promote Swedish popular music on a larger scale, both domestically and internationally. Following the success of "Främling", in 1984, Mariann introduced a new musical group, the boy band Herreys. They won both the Swedish and the International Eurovision Song Contest in Luxembourg, with their hit single "Diggi- Loo Diggi-Ley", written by Torgny Söderberg, music, and Britt Lindeborg, lyrics. Mariann had over 250 songs on the Swedish radio charts program Svensktoppen and achieved over 100 million record sales (Rylander and Karlsson, 2016).

During the late 1980s, Bert Karlsson, the successful Icebreaker, sought new challenges and opportunities. He focused on multiple projects and businesses while some of the artists, composers, and producers in his Cluster aimed to start their own ventures. In the 1990s, Bert Karlsson and Mariann collaborated to create a TV show called Fame Factory, which aired on the commercial broadcaster TV3 from 2002 to 2005. In 2006, Bert Karlsson, who had been a prominent personality in the Swedish music industry for more than three decades, decided to sell Mariann Records to Warner Music Group.

Carola, who performed "Främling", left a significant impact for various reasons. At the age of 16, she projected a blend of vulnerability, determination, integrity, self-assurance, and confidence. Her charisma was effortlessly captivating, and she was hard-working and genuine in her artistic expression. She had excellent singing skills and interpreted the song brilliantly while executing choreographic moves flawlessly. Her deep religious beliefs were evident, and during the winner's interview, she even brought the Holy Bible on stage. Her persona was a fascinating mix of the spiritual and the profane. The stage was solely focused on her, with no other distractions, as she appeared in a bright yellow outfit, which stood out vividly in colour-TV. Carola's career was revived and relaunched on a new label (Rival) in 1991 with the song "Fångad av en stormvind" [Captured by a Lovestorm]. She won both the Swedish Eurovision Song Contest and the International final in Rome.

3. Conclusions

The success of "Främling" can be attributed to a group of dedicated and skilled individuals who worked innovatively in close coordination. This team comprised experts from various fields committed to achieving their common goals. The team focused on producing a large quantity of high-quality music. Songwriters and music producers worked tirelessly in the studio, often for long hours, to create new songs. These songs were tested on different artists and released on various platforms. The study clearly indicates that facilitators, or Icebreakers, such as Bert Karlsson can play a vital role in bringing talented individuals together and opening up opportunities. The study also indicates that the "Främling" production team, or Cluster, collaborated successfully due to their prior experience and well-defined roles. The composer, Lasse Holm, used his broad musical background and previous experience in the Swedish Eurovision Song Contest to create the original idea with a catchy melody that easily stays with the listeners. The song also has an intriguing tension between the verse and chorus due to the alternation between major and minor keys and it has a catchy rhythmic basis that combines elements of both march and disco music. The lyricist, Monica Forsberg, based the text on Lasse Holm's initial idea and wrote the lyrics in a manner that allowed different listeners to interpret the meaning of the lyrics slightly differently. But the most significant factor, Carola, who was only 16 years old then, delivered an outstanding performance. Her unique singing style, which appealed to a diverse audience, was an instant hit with the listeners.

The timing for Carola in 1983 was also perfect. In 1982, Germany won the Eurovision Song Contest with "Ein bißchen Frieden", sung by 17-year-old Nicole, a high school student. It was a sensational success. This paved the way for Carola, who was even younger than Nicole. But Carola appeared and behaved in a way that was very different from Nicole. Carola sang enigmatic lyrics with erotic undertones while portraying herself as a rather innocent girl. The fact that she also held the Bible in her hand during the contest's voting procedure reinforced this image. This created a dynamic that many listeners found intriguing. However, the English translation of the text was not as ambiguous as the Swedish original, and much of the aesthetic tension present in the Swedish version was lost in translation. Another important aspect is that the song "Främling" itself was in many ways a wild card. It was another song, "Gubben i lådan", that was supposed to be the flagship for Mariann in 1983. But it was replaced by "Främling" at a late stage. This reminds us of how ABBA nine years earlier at a late stage also chose to change the song, from "Hasta Mañana" to "Waterloo", the song that ABBA won both the Swedish and international Eurovision Song Contest with in 1974 (Gradvall, 2023; Palm, 2008).

Finally, "Främling" was a big success, above all in Sweden but also in the other Nordic countries and contributed strongly to Carola's career. Her first album became one of the top-selling records ever in Sweden, with over 250,000 copies sold in the first week, earning her both gold and platinum certifications. Bert Karlsson and Mariann's innovative business model worked well to produce and promote the song. The model had clear similarities to successful music industry models in the US, pioneered by entrepreneurs like Barry Gordy Jr and Phil Spector. However, Bert Karlsson was completely unaware of this at the time. Bert Karlsson and Mariann's model was refined through a process of trial and error, leading to the success of "Främling" and many other hits in the years to come.

References

Alrup, J. and Andström, B. (eds.) (1983). *Carola: flickan som blev stjärna över en natt!*. Semic.
Aristotle (2011). *Aristotle's Nicomachean ethics.* University of Chicago Press.
Arvidsson, K. (2007). *Skivbolag i Sverige: musikföretagandets 100-åriga institutionalisering.* Göteborgs universitet.
Billboard (2023). THE TOP 30: 5. Daniel Ek, CEO Spotify. *The Billboard 2023 Power 100 List Revealed: Billboard's annual list of the music industry's top players reflects a year of change — and intensifying competition — with sharpened focus on the leaders in charge.* www.billboard.com/h/billboard-2023-power-100-executives-list/.
Brolinson, P.-E. and Larsen, H. (1981). *Rock – aspekter på industri, elektronik & sound.* Esselte studium.
Ekström, A. and Koljonen, J. (2012). *Främling: en bok om Carola.* Weyler.
Forsberg, M. and Holm, L. (1983). *Främling. EMI Music Publishing Scandinavia AB.*
Forsberg, M., Holm, L. and Slater, S. (1983). *Främling. EMI Music Publishing Scandinavia AB.*
Gradvall, J. (2023). *Vemod undercover: boken om ABBA.* [Malmö]: Inläst för Myndigheten för tillgängliga medier, MTM.

Gullö, J.-O. (2023). *Interview with Bert Karlsson*. Ursand Resort & Camping, Vänersborg. 29th of June 2023.

Gullö, J.-O., Holgersson, P.-H., Thyrén, D. and Florén, T. (2022). Icebreakers and clusters within the Swedish music wonder. Innovation in Music 2022 – Book of Abstracts: Conference Programme 17–19 June 2022, 1. https://urn.kb.se/resolve?urn=urn:nbn:se:kmh:diva-4739.

Gullö, J-O. and Thyrén, D. (2018). *Interview with Ola Håkansson*. Ten Music Group. Stockholm. 12th of December 2018.

Gullö, J-O. and Thyrén, D. (2019). Music production in Swedish higher education. History and future challenges. *Swedish Journal of Music Research*. vol. 101.

Hedlund, O. (1983). *Stikkan – Den börsnoterade refrängsångaren*. Sweden Music Förlags AB.

Holm, L. and Lundin, C. (2015). *En hel massa Lasse*. Bokfabriken.

Karlsson, B. (2006). *Mitt liv som Bert*. Sportförlaget i Europa AB.

Löfström, E. and Eklöf, J. (2019). Avsnitt 94, refuserade bidrag del II: Skara-pop. *Schlagervännerna*. Podcast. https://open.spotify.com/episode/68YqUsStgpqxJ0iuPItbqo?si=dc10e993b4c24d2d.

Ostman, L. (2018). *Hur västvärlden fylldes med musik: människorna, organisationerna och musikens kedjor*. Kulturhistoriska Bokförlaget.

Palm, C. M. (2008). *ABBA The Story – Berättelsen om supergruppen*. Wahlström & Widstrand.

Rylander, J. and Karlsson, B. (2016). *Bert Karlsson – Så blir du miljonär*. Sportförlaget i Europa AB.

Thyrén, D. Gullö, J-O. and Schyborger, P. (2020). The Denniz PoP Model: Core Leadership Skills in Music Production as Learning Outcomes in Higher Education. *Music & Entertainment Industry Educators Association*. DOI: 10.25101/20.32.

INDEX

Printed in the United States
by Baker & Taylor Publisher Services